Cambridge Tracts in Mathematics
and Mathematical Physics

GENERAL EDITORS
F. SMITHIES, Ph.D. and J. A. TODD, F.R.S.

No. 56

THE THEORY OF CLUSTER SETS

THE
THEORY OF CLUSTER SETS

BY

E. F. COLLINGWOOD, F.R.S.

AND

A. J. LOHWATER

CAMBRIDGE
AT THE UNIVERSITY PRESS
1966

PUBLISHED BY THE PRESS SYNDICATE OF THE UNIVERSITY OF CAMBRIDGE
The Pitt Building, Trumpington Street, Cambridge, United Kingdom

CAMBRIDGE UNIVERSITY PRESS
The Edinburgh Building, Cambridge CB2 2RU, UK
40 West 20th Street, New York NY 10011–4211, USA
477 Williamstown Road, Port Melbourne, VIC 3207, Australia
Ruiz de Alarcón 13, 28014 Madrid, Spain
Dock House, The Waterfront, Cape Town 8001, South Africa

http://www.cambridge.org

First published 1966
First paperback edition 2004

A catalogue record for this book is available from the British Library

Library of Congress Catalogue Card Number: 66-18115

ISBN 0 521 04695 5 hardback
ISBN 0 521 60481 8 paperback

CONTENTS

Chapter 5. CLUSTER SETS OF FUNCTIONS
ANALYTIC IN THE UNIT DISC

Chapter 6. BOUNDARY THEORY IN THE LARGE

Chapter 7. BOUNDARY THEORY IN THE SMALL

Chapter 8. FURTHER BOUNDARY PROPERTIES OF
FUNCTIONS MEROMORPHIC IN THE DISC.
CLASSIFICATION OF SINGULARITIES

Chapter 9. PRIME ENDS

PREFACE

This book is intended as an introduction to the theory of cluster sets, and as such does not claim to be comprehensive. Thus, we have had to be selective and omit some things we should have liked to include, but the space available in a Cambridge Tract is limited, and the theory continues to grow.

The founder of the theory was Painlevé who, in 1895, first gave a name—that of *domaine d'indétermination*—and the status of a distinct mathematical notion to the set of limit points of a function at a boundary point of its domain of definition. This set is now called the *cluster set* of the function at the point in question. Although it arose in connexion with analytic functions the notion is applicable to more general mappings and situations and gives rise to a theory which is essentially topological. Applied, as it has been for the most part, to problems in the theory of functions, it forms a chapter in the expanding field of Topological Function Theory. While this is our main theme and point of view, we have tried not to lose sight of the more general background.

When this book was planned there was no other on the subject available, and the need for an organized account seemed obvious. However, K. Noshiro's book *Cluster Sets* in the 'Ergebnisse' series appeared at about the time when the first draft of our book was finished. Far from regretting this, we regarded it as a fortunate coincidence since the objectives of the two series in which they appear are rather different and in consequence the two books would, to a considerable extent, complement one another. Since then, other heavy commitments have delayed completion and in the interval the theory has developed considerably. We have, therefore, had to introduce new material and reorganize parts of what we had written, but without attempting to cover the whole of the new territory recently opened up.

Although a general familiarity with classical function theory is assumed, we have included for completeness and ease of

reference in Chapters 2 and 3 most of the distinctive function-
theoretic material required in later chapters. The only other
body of distinctive material required is some elementary set-
theoretic topology which will be familiar to every analyst.

Since the interest of a theory rests ultimately on its applica-
tions, a word should perhaps be said about our final chapter.
In this we give a fairly extended treatment of the prime ends of
a plane domain by the combined cluster-set and function-
theoretic methods which have recently yielded solutions of some
long-standing problems in this subject. As well as providing an
illustration of the power of the methods now available, this
chapter, we hope, may be useful in itself.

Our bibliography is confined principally to those writings to
which we actually refer and is far from complete. But references
omitted by us will generally be found in Noshiro's bibliography,
where the coverage up to the date of its publication is much
wider; we have, however, added references to the most recent
literature.

As regards the arrangement of the book Chapter 1 is intro-
ductory, Chapter 2 contains a brief expository account of the
classical theory of radial limit values, with some additional
material on Blaschke products. In Chapter 3 we give results
from the theory of conformal mapping which will be needed in
later chapters. The experienced reader may proceed directly to
Chapter 4 where the intrinsic properties of cluster sets of func-
tions defined in the unit disc, but not necessarily analytic, are
introduced. In Chapter 5 we study the boundary behaviour of
an important class of functions, Seidel's Class U, which are
analytic in the disc, together with the Gross–Iversen Theorem
on exceptional values and its generalizations. In Chapter 6 we
study the global cluster set of a function meromorphic in the
disc and its relations with the global exceptional and asymptotic
sets. This we call boundary theory in the large. Chapter 7 is on
boundary theory in the small; in it we apply the same techniques
to the local cluster set, that is to say, the cluster set at a point of
the unit circumference, and its relations with the local excep-
tional and asymptotic sets. The main results of these two
chapters are the appropriate analogues of Iversen's Theorem 1.6

for the disc in the large and in the small respectively. In Chapter 8 we have collected a number of recent results on the classification of the singularities of functions meromorphic in the disc and their distribution on the circumference, together with uniqueness and existence theorems under given boundary conditions. Chapter 9, which concludes the book, is on boundary correspondence and prime ends.

We are grateful to various friends, who have seen parts of the book at different stages, for much helpful criticism and in particular to F. W. Gehring and G. Piranian. We are also much indebted for the suggestions and comments of the referee and our editors. We cannot hope to have avoided all the pitfalls but we should certainly have fallen into more of them had it not been for their help.

<div style="text-align:right">

E.F.C.
A.J.L.

</div>

INTRODUCTION

1. The cluster set. Definition

The notion of a cluster set was first formulated explicitly by Painlevé ([1], p. 438) in his well-known Stockholm lectures of 1895 on differential equations. Painlevé introduced the cluster set, which he called *domaine d'indétermination*, as a descriptive notion to characterize in an intuitive way the behaviour of an analytic function in the neighbourhood of a singularity in terms of the properties of the set of all its limits at the singularity, and to classify the singularities of a function in terms of these cluster sets.

Although Painlevé was considering only analytic functions, the notion of a cluster set does not depend, any more than does the notion of a limit, upon the function being analytic or even continuous; nor need it be restricted to complex-valued functions of a complex variable. It is equally applicable to real functions of a real variable and to mappings in spaces of higher dimensions, as well as to general topological spaces, although not necessarily equally fruitfully in all cases. We shall therefore introduce the definitions in a sufficiently general form to include the special cases we shall require in later chapters.

Let the function $w = f(z)$ be defined in a domain D which it maps into the w-sphere S. Unless otherwise stated, we shall take D to be in the z-plane, but if D is an open set on the real line, we shall assume the mapping to be into the real line or into a great circle in S. D is connected, and, initially, we impose no restriction on its connectivity, which may be finite or infinite. The boundary of D is denoted by $\mathrm{Fr}\,D$, while $\overline{D} = D \cup \mathrm{Fr}\,D$ is the closure of D.

Given any point z_0 of \overline{D}, we may define the *cluster set* $C_D(f, z_0)$ of $f(z)$ at z_0 in either of the following two equivalent ways:

(i) $C_D(f, z_0)$ is the set of points α of the w-sphere S (or real line if f is real) such that there exists a sequence $\{z_n\} \subset D - z_0$ such that $\lim_{n \to \infty} z_n = z_0$ and $\lim_{n \to \infty} f(z_n) = \alpha$. Here $D - z_0$ denotes the

deleted domain obtained by removing z_0 from D. In general, however, we shall be concerned with the case when $z_0 \in \text{Fr } D$, in which case $D - z_0 = D$.

(ii) $C_D(f, z_0) = \bigcap\limits_{r>0} \overline{D_r}$, where $D_r = f(\delta_r \cap (D - z_0))$ and δ_r is the disc $|z - z_0| < r$.

It follows immediately from the definition in either form that $C_D(f, z_0)$ *is a non-empty closed set.* A cluster set which contains only a single point will be called *degenerate*, while if it contains more than one point, it will be called *total* or *subtotal* according as it does or does not cover the entire w-sphere S, or, in the case where f is a real function, the entire line.

The question whether or not $C_D(f, z_0)$ is connected depends both upon D and f. If, for some $r > 0$, δ_r contains no point of $\text{Fr } D - z_0$ (if $z_0 \in \text{Fr } D$), then δ_r is a neighbourhood of z_0; but if, for all $r > 0$, δ_r contains points of $\text{Fr } D - z_0$ and there is a component δ_r^* of $\delta_r \cap D$ with z_0 in its frontier, we say that δ_r^* is a *relative neighbourhood* of z_0. *If $f(z)$ is continuous in D, and if D is locally connected at z_0, the relative neighbourhoods are connected for all small $r > 0$.* Hence D_r is connected, and it follows from (ii) that $C_D(f, z_0)$ *is either a continuum or a point.*

The theory has been built up round three situations which are geometrically the simplest: namely (a) the case where z_0 is an isolated frontier point or is an interior point of D; (b) the case where D is the unit disc and z_0 a frontier point; and (c) the case where $\text{Fr } D$ is totally disconnected and $z_0 \in \text{Fr } D$. The situations (a) and (b) are, in a sense, extremal, and (c) intermediate. The problems presented by (b) are, in general, much more complicated than those presented by (a) and, to a lesser extent, those presented by (c). It is very largely with problems concerning (b), where f is defined in a disc, that we shall be dealing in this book.

2. Definition of a partial cluster set

Let \mathscr{G} be any infinite subset of D on the real line or in the z-plane, and let $z_0 \in \mathscr{G}'$, where \mathscr{G}' is the derived set of \mathscr{G}. Then, exactly as in §1, we define the cluster set $C_{\mathscr{G}}(f, z_0)$ of $f(z)$ on \mathscr{G} by

$$C_{\mathscr{G}}(f, z_0) = \bigcap_{r>0} \overline{D_r(\mathscr{G})} \subset C_D(f, z_0),$$

where $D_r(\mathscr{G}) = f(\delta_r \cap (\mathscr{G}-z_0))$ is the image of $\delta_r \cap (\mathscr{G}-z_0)$ on S. $C_D(f, z_0)$ is the *complete cluster of* $f(z)$ *at* z_0 and will be written simply as $C(f, z_0)$ if there is no risk of ambiguity. $C_{\mathscr{G}}(f, z_0)$ is a *partial cluster set of* $f(z)$ *at* z_0. We shall find, however, that in a number of important cases a partial cluster set is not a proper subset of the complete cluster set.

If $z_0 \in \text{Fr}\, D$, where $\text{Fr}\, D$ is suitably restricted, the cluster set of f at z_0 by approach on the boundary, i.e. the *boundary cluster set*, is defined by means of the cluster sets at points of $\text{Fr}\, D$ in place of the values at such points, which are, in general, undefined. We shall give precise definitions as they are required.

By the argument of the previous paragraph we have

THEOREM 1.1. *If* $f(z)$ *is continuous on a set* \mathscr{G} *which is locally connected at a point* $z_0 \in \overline{\mathscr{G}}$, *the cluster set* $C_{\mathscr{G}}(f, z_0)$ *is either a continuum or a point.*

3. Illustrations

The notion of a cluster set may be illustrated by trivial examples. Thus, if $z_0 = 0$ and $f(z) = f(r\,e^{i\theta}) = 1$ for all points for which $r > 0$ and θ are both rational, while for all other points $f(z) = -1$, the cluster set $C(f, z_0)$ consists of the points $w = 1$ and $w = -1$. For the function $f(z) = e^{1/z}$ the set $C(f, 0)$ is total; and, to take a third example, if $f(z) = e^{i/|z|}$, then $C(f, 0)$ is the circumference $|w| = 1$ and is thus non-degenerate but subtotal. The real function $y = f(x) = \sin 1/x$ has the cluster set $C(f, 0)$ consisting of the real segment $-1 \leqslant y \leqslant 1$ whose endpoints ± 1 are the upper and lower limits of $\sin 1/x$ at $x = 0$ from both directions.

For a cluster set to be a continuum it is not necessary for the set \mathscr{G} on which it is generated to be connected, or for the mapping function f to be continuous on \mathscr{G}. It is easy to construct a real function which exists and is totally discontinuous on a totally disconnected set dense on a segment of the real line, whose cluster set is a closed segment. Let $f_0(x) = x$ for all rational values of x in $0 < x < 1$ having even denominators in their lowest terms while $f_0(1) = 1$ and $f_0(x) = 0$ for all other rational values of x in the

interval. We now define $f(x)$ on the totally disconnected set \mathscr{G} of rationals in $0 < x \leqslant 1$ as follows:

$$
\begin{aligned}
f(x) &= f_0(2(x - \tfrac{1}{2})) \quad \text{for} \quad \tfrac{1}{2} < x \leqslant 1 \\
&= f_0(4(x - \tfrac{1}{4})) \quad \text{for} \quad \tfrac{1}{4} < x \leqslant \tfrac{1}{2} \\
&\quad\cdots\cdots\cdots\cdots\cdots\cdots\cdots\cdots\cdots\cdots\cdots\cdots\cdots \\
&= f_0(2^n(x - 2^{-n})) \quad \text{for} \quad 2^{-n} < x \leqslant 2^{-n+1} \\
&\quad\cdots\cdots\cdots\cdots\cdots\cdots\cdots\cdots\cdots\cdots\cdots\cdots\cdots\cdots\cdots\cdots
\end{aligned}
$$

The notion of cluster set is in fact a very general one which is applicable to a wide range of topological mappings. Historically, the theory has developed within the theory of functions, to which it has added a number of striking theorems; for the most part, the approach to the theory of cluster sets has been function-theoretic, since the results were to be applied directly in the theory of functions. The more general approach, with the hypotheses less deeply implanted in the theory of functions, is a comparatively recent development from which further progress may be expected. If the subject were to be treated in a strictly logical fashion, one would start from the most general topological ideas and would then particularize whenever a more specialized result is wanted. At the present stage of development of the theory, however, we prefer to take our point of departure in discussing the various divisions of the theory from within the framework of familiar ideas belonging to continuous mappings, or to the special case of analytic functions, and to introduce generalizations in their appropriate contexts as they arise.

4. Theorems of Weierstrass and Painlevé

The earliest developments in the theory of cluster sets were concerned with the behaviour of an analytic function in the neighbourhood of either an isolated essential singularity or else a discontinuous set of singularities. Indeed, the earliest result which may be said to belong properly to the theory of cluster sets is the following classical theorem, commonly attributed to Weierstrass [1],† which we state in the terminology of cluster sets.

† The date of Weierstrass's paper was 1876, but the theorem bearing his name had, in fact, first been proved by Briot and Bouquet [1], p. 39, in the *first edition* of their book on elliptic functions in 1859, and a proof was also given by

THEOREM 1.2. *Let z_0 be an isolated point of the set E, $E \subseteq D$. If $f(z)$ is meromorphic in $D - E$, then $C(f, z_0)$ is either total or degenerate.*

It is natural to ask whether the condition on the set E in the statement of Theorem 1.2 can be relaxed. Painlevé broached the question in his Stockholm lectures, but it was not until ten years later† that he proved that the same conclusion holds if E is of linear measure zero and z_0 is any point of E, a result which was later rediscovered by Besicovitch [1] who also showed that the theorem fails if E is an arbitrary set of positive linear measure.

5. Proofs of the theorems of Painlevé and Besicovitch

In the usual terminology, we shall say that a set E in the plane is of *linear measure zero* if E can be covered by a sequence of circles the sum of whose diameters can be made arbitrarily small. We now prove

THEOREM 1.3. *Let the set E, $E \subset D$, where D is a domain, be of linear measure zero, and let $f(z)$ be meromorphic in $D - E$. Then for each point $z_0 \in E$, the cluster set $C(f, z_0)$ is either total or degenerate.*

We shall actually prove more, namely, that at points z_0 for which $C(f, z_0)$ is degenerate, the function $f(z)$ is meromorphic. We prove first the following lemma on removable singularities.

LEMMA. *Let $f(z)$ be analytic and bounded in a simply connected domain D except at a closed set E in D of linear measure zero. Then $f(z)$ may be extended to points of E so as to be analytic in all of D.*

Casorati [1], p. 454, in 1868. It is true that the enunciation of Briot and Bouquet was imprecise since they used the word *acquiert* for approach to a limit, but their proof was correct and was in fact the classical one given later both by Casorati and Weierstrass. The theorem does not, however, appear in the second edition (1875) of their book, although the section in which it had appeared in the first edition was otherwise unaltered. In the Soviet Union the theorem is frequently referred to as Sokhotskii's theorem on the basis of his dissertation (St Petersburg, 1873); see, for example, Markushevich [1], p. 189, or Privalov [2], p. 190 (or [3], p. 136).

† Painlevé's proof of Theorem 1.2, which was substantially the same as that given here, was first published, with due acknowledgement, by Zoretti [1]. The proof is reproduced in Zoretti [2], which also contains a fairly complete account of the theory of cluster sets as developed by Painlevé and his pupils. See also Pompéiu [1].

We may form a simple, closed, rectifiable curve Γ lying in D and enclosing the set E. The function

$$F(z) = \frac{1}{2\pi i} \int_\Gamma \frac{f(\zeta)}{\zeta - z} d\zeta \qquad (1.5.1)$$

is analytic and bounded in the domain D_1 enclosed by Γ. Since E is of linear measure zero, it is possible to cover E with a sequence of circles whose perimeters have an arbitrarily small total length. For an arbitrary but fixed z in $D_1 - E$ and for any sufficiently small $\epsilon > 0$, we may, since E is closed, apply the Borel covering theorem and enclose E by a finite number of mutually disjoint rectifiable curves $\gamma_1, \gamma_2, ..., \gamma_n$ having total length less than ϵ and lying in D_1 in such a way that z lies outside each of the γ_k. Now in that part of D_1 lying outside the γ_k, $f(z)$ is analytic and may be represented by means of the Cauchy integral formula

$$f(z) = \frac{1}{2\pi i} \int_\Gamma \frac{f(\zeta)}{\zeta - z} d\zeta + \sum_{k=1}^n \frac{1}{2\pi i} \int_{\gamma_k} \frac{f(\zeta)}{\zeta - z} d\zeta$$

$$= F(z) + \sum_{k=1}^n \frac{1}{2\pi i} \int_{\gamma_k} \frac{f(\zeta)}{\zeta - z} d\zeta, \qquad (1.5.2)$$

whence $\qquad |f(z) - F(z)| \leqslant \frac{1}{2\pi} \sum_{k=1}^n \left| \int_{\gamma_k} \frac{f(\zeta)}{\zeta - z} d\zeta \right| < \frac{\epsilon}{2\pi} \frac{M}{\kappa}, \qquad (1.5.3)$

where M is the bound on $|f(z)|$ in D and κ is the minimum distance from z to the set $\Gamma \cup \bigcup_{k=1}^n \gamma_k$. Since E is at a positive distance δ from Γ, we can, for any $\epsilon_1 < \epsilon$, enclose E in a finite set $\Gamma_1 = \bigcup_{j=1}^m \gamma_j'$ of disjoint rectifiable curves having total length less than $\min(\delta, \epsilon_1)$, and therefore contained in Γ, so that (1.5.3) holds with ϵ_1 in place of ϵ. Thus we see that $F(z)$ coincides with $f(z)$ in $D_1 - E$, so that the cluster set $C(f, z_0) = C(F, z_0) = F(z_0)$ is degenerate for all $z_0 \in E$, since $F(z)$ is analytic in D_1. Putting $f(z) = F(z)$ for $z \in E$, we have the required extension, and the lemma is proved.

To prove Theorem 1.3 we may assume that E is the set of singularities of $f(z)$ in D and therefore closed. For if $z_0 \in E$ is not a singularity, then $C(f, z_0) = f(z_0)$ is degenerate. Suppose then that $C(f, z_0)$ is subtotal at some point $z_0 \in E$. There is then a number c

and a neighbourhood d of z_0 such that $\phi(z) = [f(z) - c]^{-1}$ is analytic and bounded in $d - E_1$, where E_1 is the closed subset of E contained in d. It now follows from the lemma that the cluster set $C(\phi, z_0)$, and therefore also $C(f, z_0)$, is degenerate. This proves the theorem.

A point $z_0 \in E$ for which $C(f, z_0)$ is degenerate is a removable singularity, while the points of E at which the cluster set is total are essential singularities. More generally, we call any point at which the cluster set of a uniform function is non-degenerate an essential singularity of the function.

The range and the asymptotic set

6. Definitions. Theorems of Picard, Cartwright and Iversen

In addition to the concept of the cluster set there are two other notions which have played a central rôle in the theory of the boundary behaviour of meromorphic functions. These are the *range* of values assumed arbitrarily near the boundary and the *asymptotic values* approached along continuous paths to the boundary. Indeed, it is on the interrelations between these sets and the complete and partial cluster sets at a point or boundary that much of the modern theory of cluster sets is founded.

Given a function $f(z)$ defined in some deleted neighbourhood $D - z_0$ of the point z_0 in the complex plane or on the real line we define the *range* $R(f, z_0)$ as the set of points α on the w-sphere such that there exists a sequence $\{z_n\} \subset D - z_0$ for which $\lim\limits_{n \to \infty} z_n = z_0$ and $f(z_n) = \alpha$ for all n. In the notation of §1 we have, alternatively,
$$R(f, z_0) = \bigcap_{r>0} D_r.$$

Unlike the cluster set, the range $R(f, z_0)$ may be empty, as it is, for example, for $f(z) = (z - z_0)^N$ or for any $f(z)$ of finite valency in $D - z_0$. Similarly, the range $R_{\mathscr{G}}(f, z_0)$ on \mathscr{G} of $f(z)$ at $z_0 \in \mathscr{G}'$ is defined as the set of points α such that there exists a sequence $\{z_n\} \subset \mathscr{G}$, $\lim\limits_{n \to \infty} z_n = z_0, f(z_n) = \alpha$; or, alternatively,
$$R_{\mathscr{G}}(f, z_0) = \bigcap D_r(\mathscr{G}),$$
where now $D_r(\mathscr{G}) = f(\delta_r \cap (\mathscr{G} - z_0))$.

By definition, a value α is an *asymptotic value* of $f(z)$ at z_0 and belongs to the *asymptotic set* of such values, denoted by $\Gamma(f, z_0)$, if there is a continuous curve $z = z(t)$, $0 < t < 1$, such that $z(t) \subset D - z_0$, $\lim_{t \to 1} z(t) = z_0$ and $\lim_{t \to 1} f(z(t)) = \alpha$. Similarly, we define $\Gamma_{\mathscr{G}}(f, z_0)$ by imposing the condition $z(t) \subset \mathscr{G} - z_0$ where now \mathscr{G} must be a connected domain; but if \mathscr{G} is in fact the domain of definition of $f(z)$ near z_0 we shall, as in the case of the cluster set and range, write simply $\Gamma(f, z_0)$.

As in the case of the cluster set the earlier developments with the sets $R(f, z_0)$ and $\Gamma(f, z_0)$ were linked with the assumption that z_0 was an isolated boundary point and that $f(z)$ was defined in $D - z_0$. The first result in this group was the now famous theorem of Picard [1] which may be expressed in the form

THEOREM 1.4. *Let z_0 be an isolated point of the set $E \subset D$. If $f(z)$ is meromorphic in $D - E$ and E is the set of its essential singularities, then the complement $\mathscr{C}R(f, z_0)$ of $R(f, z_0)$ consists of at most two points.*

While it is easy to extend Theorem 1.4 to the case of a point z_0 which is the limit point of isolated singular points, it is not so simple to draw the corresponding analogue to Theorem 1.3 for the range. The set $\mathscr{C}R(f, z_0)$ can contain many more than two points under the hypotheses of Theorem 1.3, as we shall indicate in § 8, but the following result of Cartwright [2]† shows that under the hypotheses of Theorem 1.3, we can expect a certain restriction on the size of $\mathscr{C}R(f, z_0)$.

THEOREM 1.5. *If the set E of singularities of a meromorphic function is of linear measure zero, then $\mathscr{C}R(f, z_0)$ is of plane measure zero for all z_0 in E.*

The first classical theorem concerning the set $\Gamma(f, z_0)$ was given relatively late, compared to the theorems of Weierstrass and Picard, and, like these theorems, was restricted to isolated singularities.

THEOREM 1.6.‡ *If $z_0 \epsilon E$ is an isolated essential singularity of $f(z)$, then $\mathscr{C}R(f, z_0) \subset \Gamma(f, z_0)$.*

† It is an easy deduction from Theorem 1.3 that $R(f, z_0)$ is everywhere dense; see also Theorem 4.2.

‡ Iversen [1].

If z_0 is not an isolated singularity of $f(z)$, but is a limit point of isolated singularities, then it is trivial to show that $\mathscr{C}R(f, z_0)$ *contains at most two points and*

$$\mathscr{C}R(f, z_0) \subset \bigcap_{z \in d \cap E - z_0} \Gamma(f, z),$$

where d is any neighbourhood of z_0. Before we discuss further relations among the sets $C(f, z_0)$, $R(f, z_0)$, and $\Gamma(f, z_0)$, we shall find it convenient to define the *capacity* of a set E.

7. The notion of capacity

Let $\mu(E)$ be a non-negative, additive set function defined on all Borel sets E in the plane. Let F be a closed bounded set in the plane having a connected complement G, and let \mathscr{M} be the set of all set functions μ with the property

$$\int_{\zeta \in F} d\mu(\zeta) = 1. \tag{1.7.1}$$

A set function μ with the property (1.7.1) will be called a *distribution of unit mass* on F. The function

$$u(z) = \int_{\zeta \in F} \log \frac{1}{|z - \zeta|} d\mu(\zeta)$$

is harmonic in G; we consider first the $\sup_{z \in G} u(z)$, which may be infinite, and then the quantity

$$V_F = \inf_{\mu \in \mathscr{M}} (\sup_{z \in G} u(z)), \tag{1.7.2}$$

which we call the *equilibrium potential* of the set F. If the set is so thin that a unit mass (or charge) cannot be distributed on it without inducing an unbounded potential, we say that F is of *capacity zero*; the capacity of the set F is defined to be

$$\operatorname{cap} F = e^{-V_F}. \tag{1.7.3}$$

The capacity of a Borel set E is then defined as the supremum of the capacities of all closed subsets of E, i.e.

$$\operatorname{cap} E = \sup_{F \subset E} (\operatorname{cap} F). \tag{1.7.4}$$

Thus it follows that $A \subseteq B$ implies $\operatorname{cap} A \leqslant \operatorname{cap} B$. Also, from the definition, $\operatorname{cap} A = \operatorname{cap} B = 0$ implies $\operatorname{cap}(A \cup B) = 0$. For,

otherwise, there is a positive mass distribution on $A \cup B$ which induces a bounded potential on this set. But, under this distribution, at least one of the sets A, B carries a positive mass which induces an unbounded potential. Therefore the union of a countable set of sets of capacity zero is of capacity zero. It follows at once that the capacity of a countable set is also zero. A set which is of capacity zero is of linear measure zero, but not conversely; the middle-thirds set of Cantor, for example, is of positive capacity.† The notion of capacity has played an important rôle in the theory of functions, where a set of capacity zero has proved to be a rather natural exceptional set, and, as we shall see in some of the developments of the later chapters, the notion is important also in the theory of cluster sets. We shall find the following property of closed sets of capacity zero of importance in the sequel:

THEOREM 1.7. *If E is a closed, bounded set of capacity zero, then there exists a distribution μ of unit mass on E with the property that the potential*

$$u(z) = \int_E \log \frac{1}{|z - \zeta|} \, d\mu(\zeta)$$

tends to $+\infty$ as z tends to an arbitrary point of E.

The function $u(z)$ is sometimes called the Evans potential or the Selberg potential‡ of the set E.

8. Counter-example for non-isolated singularities

We shall use the notion of capacity to show that, for non-isolated singularities, a direct analogue of Theorem 1.4 is not generally valid for the range set $R(f, z_0)$ under the conditions of Theorem 1.3. Let E_w be a perfect, nowhere dense set of capacity zero lying in $|w| < 1$; it follows from the fact that E_w is of capacity zero that the complement of E_w in $|w| < 1$ is an open,

† We shall give no proofs of the results described in §7; we refer the reader to Nevanlinna [1] and Frostman [1].

‡ See, for example, Selberg [1], who proves this result by means of the fact that a closed set E not separating the plane is of positive capacity if and only if the complement of E has a Green's function. Noshiro [2] gives a simple proof of this result using the equivalence of the notions of capacity and transfinite diameter.

connected set or domain G. The universal covering surface† R_G of G is simply connected and may be mapped conformally onto $|z| < 1$; if $w = f(z)$ denotes the inverse function of the mapping function, it follows that $f(z)$ is analytic and bounded, $|f(z)| < 1$, in $|z| < 1$ and assumes infinitely often in $|z| < 1$ all values of G, but no values of E_w. Under the mapping function the circumference $|w| = 1$ corresponds to a countably infinite set of open arcs $\{I_n\}$ on $|z| = 1$, while the image of E_w on $|z| = 1$ is a perfect, nowhere dense set E *of positive capacity but of linear measure zero*.‡ Since the function $f(z)$ is continuous with modulus 1 on each arc I_n, it follows that $f(z)$ may be continued analytically beyond the arcs I_n by means of the reflexion principle $f(1/\bar{z}) = 1/\overline{f(z)}$ to give a function $F(z)$ which is analytic and uniform everywhere in the plane except at the set E, which forms a set of essential singularities of $F(z)$. From Theorem 1.3 it follows that $C(F, z_0)$ is total for each $z_0 \in E$, but, by the construction, the set $\mathscr{C}R(F, z_0)$ must contain the non-countable set E_w for each $z_0 \in E$.

9. Case of thin sets of non-isolated singularities

For a set E of singularities of capacity zero a theorem due to af Hällström [1], stronger than Theorem 1.5, holds, namely:

THEOREM 1.8. *If the set E of essential singularities of a meromorphic function $f(z)$ is of capacity zero, then $\mathscr{C}R(f, z_0)$ is of capacity zero for all $z_0 \in E$.*

An outline of the proof is as follows. Since E is of capacity zero it is also of linear measure zero, so that for $z_0 \in E$, $C(f, z_0)$ is total. It follows that $R(f, z_0)$ is everywhere dense.§ We may assume without loss of generality that $\infty \in R(f, z_0)$. Let ρ be any positive number and let C be a circular disc $|z - z_0| < \rho$. Describe a simple closed curve γ surrounding z_0 in $C - E$ and denote by δ_ρ the domain enclosed by γ from which E is deleted and write $\mathscr{D}_\rho = f(\delta_\rho)$. $\mathscr{C}\mathscr{D}_\rho$ is compact and contains no continuum, i.e. is

† For the concept of *universal covering surface*, see Pfluger [1] and Nevanlinna [1]; see also Springer [1].
‡ Myrberg [1]. The construction here is due to Seidel [2].
§ The proof that $R(f, z_0)$ is dense in the interior of $C(f, z_0)$ is given in Chapter 4 (Theorem 4.2).

totally disconnected. We show that $\mathscr{C}\mathscr{D}_\rho$ is of capacity zero. Otherwise, there is a non-constant bounded harmonic function $U(w)$ in \mathscr{D}_ρ. Suppose $U(w_1) \neq U(w_2)$, where w_1, w_2 are distinct points of \mathscr{D}_ρ, and consider $H(z) = U(f(z))$ in δ_ρ. But since H is harmonic and bounded in δ_ρ each point of E is a removable singularity for H in δ_ρ, contrary to hypothesis.

Also, there exist sequences $\{z_\nu^{(1)}\}$, $\{z_\nu^{(2)}\}$ in δ_ρ such that $z_\nu^{(1)} \to z_0$, $z_\nu^{(2)} \to z_0$, and $f(z_\nu^{(1)}) \to w_1$, $f(z_\nu^{(2)}) \to w_2$ so that $H(z_\nu^{(1)}) \to U(w_1)$, $H(z_\nu^{(2)}) \to w_2$ from which it follows that z_0 is not removable. Hence $\mathscr{C}\mathscr{D}_\rho$ is of capacity zero.

In the example $F(z)$ of § 8 above the set E of singularities is of *positive* capacity while $\mathscr{C}R(f, z_0)$, $z_0 \in E$, is of capacity zero. The question naturally arises whether the conclusion of Theorem 1.8, which falls very far short of the 'Picard property', namely that $\mathscr{C}R(f, z_0)$ contains at most two points, established by Theorem 1.4, can be strengthened. More specifically, what are the weakest conditions on the set E of essential singularities under which $\mathscr{C}R(f, z_0)$, $z_0 \in E$, must have the Picard property? This remains an open question. It was at one time conjectured† that the Picard property would hold if E were of capacity zero, but a recent example of Matsumoto [2] shows that this is not so. As against this, Carleson [1] has constructed an example of a closed, linear set of *positive* capacity such that no non-rational function meromorphic outside the set can omit four values. The capacity of E is therefore not by itself decisive.

Turning now to the asymptotic set $\Gamma(f, z_0)$ for points $z_0 \in E$, where E is of capacity zero, there is again an analogue, due to Cartwright [2], of Theorem 1.6, namely

THEOREM 1.9. *If the set E of essential singularities of a meromorphic function $f(z)$ is of capacity zero, then for any $\alpha \in \mathscr{C}R(f, z_0)$, $z_0 \in E$, either $\alpha \in \Gamma(f, z_0)$ or $\alpha \in \Gamma(f, z_n)$ for every z_n in a sequence of points $\{z_n\} \subset E$ such that $\lim_{n \to \infty} z_n = z_0$.*

In general terms, then, we may say that for 'thin' discontinuous sets E of essential singularities the set of exceptional values $\mathscr{C}R(f, z_0)$, $z_0 \in E$, becomes larger and the relation between

† Lohwater [7].

$\mathscr{C}R(f,z_0)$ and the asymptotic set $\Gamma(f,z_0)$ becomes looser as E grows from an isolated set to a perfect set of linear measure zero, while the basic property that $C(f,z_0)$ is total remains unchanged. Although it may be considerably elaborated, the general character of the theory of cluster sets in the case of functions having a 'thin' set of singularities is fairly well represented by the hierarchy of theorems we have stated in this chapter. If, however, the condition that E be of linear measure zero is dropped and, in particular, if E is allowed to contain a continuum, then $C(f,z_0)$, $z_0 \in E$, is no longer necessarily total and all the foregoing theorems may fail, so that we are in a different situation requiring separate study. Characteristic problems are those of the boundary behaviour of functions defined in a simply connected domain whose boundary contains more than one point and which can therefore be mapped conformally onto an open circular disc. Cluster-set theory for functions defined in a disc is thus fundamental, both as to results and methods, for the theory of cluster sets as a whole and the greater part of this book is devoted to it. Some of the material of this theory is classical; but much is comparatively recent, and the synthesis of the new with the old has considerably extended the field of application.

10. Historical note and summary

It was before the measure theories of Borel and Lebesgue had been formulated that Painlevé introduced the notion of a cluster set in his Stockholm Lectures. As we have seen, set theory without a satisfactory definition of a set of measure zero was inadequate for handling some of the problems which Painlevé had set himself. As Painlevé himself discovered, the purely topological character of the set of singularities of an analytic function does not in general determine the topological character of its cluster sets. Metrical conditions are also required, which explains the halting steps of the first few years. The proof of the Painlevé–Besicovitch theorem, replacing an incorrect assertion in the Stockholm Lectures, was first published in 1905 in a paper by Zoretti†

† See the footnote on p. 5.

which also contained other assertions, not all of them correct, on the relation between the metrical properties of the set of singularities of a function and the topological character of its cluster sets. An improved but still not entirely correct account was given in Zoretti's book [2] in 1911. Finally, Denjoy [1–5] at about the same date cleared up the doubtful points and effectively completed this part of Painlevé's theory. The other part of his theory, that of the cluster sets of the solutions of ordinary differential equations, has not been followed up, and it remains an open question whether further research in this direction would be fruitful.†

Painlevé's theory, as far as he and his pupils carried it, was essentially one of the behaviour of a meromorphic or regular function at a 'thin' boundary. The theory of cluster sets at a continuous boundary really begins with the work on radial limits of functions analytic in the unit disc dating back to Fatou's memoir [1] of 1906 and leading, during the next twenty years, to the theorems of F. and M. Riesz, Lusin and Privalov, and Plessner. This stream of work arose quite independently of Painlevé's ideas, but we can now recognize it as belonging to the theory of cluster sets, although forming a highly specialized part of it, and providing some of its essential tools. Fatou's theorem was applied in 1912 by Carathéodory [2] to the solution of the long-standing problem of the boundary correspondence between the unit disc and an arbitrary simply connected domain under conformal mapping. The notion of a prime end, which Carathéodory then introduced, is essentially a cluster-set notion, the correspondence being shown to be between the points of the unit circle and the prime ends of the domain whose *impressions* are the cluster sets of the mapping function at the corresponding points. Recent work in cluster-set theory has led to the solution of a number of outstanding problems and has brought Carathéodory's theory of prime ends to something approaching completion. Our final chapter is devoted to this theory as being one of the most striking applications of cluster sets.

The first systematic development underlying the modern

† See Painlevé [1–4], Zoretti [1, 2] and Gross [4], in which an erroneous proof in Zoretti [2] is corrected, for this theory.

theory of cluster sets was undertaken by F. Iversen between 1916 and 1922, and, independently of him, by W. Gross in 1917–18. There was considerable overlapping; but between them these authors established many of the basic properties of cluster sets of analytic functions.

However, the general theory of cluster sets remained for another ten years where Gross and Iversen had left it, and the next stage of systematic development came in 1932 and 1934 with the work of Seidel to whom the term *cluster set* is due. This was contemporary with the independent work of Beurling. Seidel's work was concerned mainly with bounded analytic functions, following an idea of Nevanlinna, and Beurling's with what we call the Iversen–Beurling Theorem.

The work of which we have been speaking was all concerned with properties of analytic functions. But, as we have already remarked, the notion of a cluster set is not restricted to this class of functions, since it is applicable to arbitrary functions. The theory of the cluster sets of arbitrary functions originated with W. H. Young [1, 2] in 1908. Young, who was clearly unaware of Painlevé's work, studied only real functions. He began by proving that the points of inequality of the right and left upper and lower limits respectively of an arbitrary single-valued real function of a single variable are countable, and, more generally, that for such a function, the points of inequality of right and left cluster sets, although not under so formal a terminology, are also countable. Young later extended his work to arbitrary real functions of several variables and showed for the first time the significance of Baire's notion of category in this field. For twenty years his work remained virtually unknown, perhaps for lack of a suitable terminology and notation to give point to the ideas. It was summarized in Young's last published paper [3] in 1928 and was taken up in the thirties by Blumberg [1], who in 1918 had independently discovered Young's first theorem on the points of asymmetry of upper and lower limits, and subsequently by others. In the last few years the extension to arbitrary complex functions and more general mappings has added unexpected and significant theorems to the general theory of cluster sets.

The modern period may be said to have begun in the middle thirties, and a considerable literature has grown up concerning many aspects of the theory, both for analytic functions and for more general mappings, including light interior and pseudo-analytic mappings as well as arbitrary functions, much of it in the last decade.

CHAPTER 2

FUNCTIONS ANALYTIC IN A CIRCULAR DISC

Theorems of Fatou and F. and M. Riesz

1. Introduction

The theorems to be developed in this chapter belong, for the most part, to an area of the theory of boundary behaviour which has become classical, namely, the theory of bounded analytic functions and functions of bounded characteristic. In the first half of the chapter we discuss Fatou's theorem and the cycle of theorems related to it. In the second half we discuss the theory of Blaschke products; and we have oriented this theory to show the extent to which theorems of the type of Fatou's theorem may be generalized as far as approach to the boundary is concerned.

2. Fatou's theorem on radial limits

The theorem, first proved by Fatou in 1906, is the following:

THEOREM 2.1. *If $f(z) = \sum_{n=0}^{\infty} a_n z^n$ is analytic and bounded, i.e.* $|f(z)| < M < \infty, for |z| < 1, then the radial limits f(e^{i\theta}) = \lim_{r \to 1} f(r e^{i\theta})$ *exist for all points $e^{i\theta}$ on $|z| = 1$, except possibly for a set of linear measure zero.*

Several proofs of this theorem are known and are set out in detail in standard texts (Carathéodory, Bieberbach, Nevanlinna, Zygmund). We therefore restrict ourselves to giving the outline of a proof based on the theory of power series which uses ideas that we shall require later on.

We show first that $\sum_{n=0}^{\infty} |a_n|^2$ is bounded. For, writing $z = r e^{i\theta}$, we have

$$|f(z)|^2 = f(z)\overline{f(z)} = \sum_{n=0}^{\infty} a_n r^n e^{in\theta} \sum_{n=0}^{\infty} \overline{a}_n r^n e^{-in\theta},$$

where the bars denote conjugates, so that, multiplying the series by the usual Cauchy rule, as we may do since both are absolutely convergent, we obtain

$$|f(z)|^2 = \sum_{n=0}^{\infty} c_n(r, \theta),$$

where

$$c_n(r, \theta) = (a_0 \bar{a}_n e^{-in\theta} + a_1 \bar{a}_{n-1} e^{-i(n-2)\theta} + \ldots + a_n \bar{a}_0 e^{in\theta}) r^n.$$

Now $\sum_{n=0}^{\infty} c_n(r, \theta)$ is uniformly convergent in $0 \leqslant \theta \leqslant 2\pi$ since both the series on the right are absolutely convergent. On integration term by term only the terms

$$\int_0^{2\pi} a_\nu \bar{a}_\nu r^{2\nu} \, d\theta = 2\pi |a_\nu|^2 r^{2\nu}$$

do not vanish so that

$$\sum_{n=0}^{\infty} |a_n|^2 r^{2n} = \frac{1}{2\pi} \int_0^{2\pi} |f(r e^{i\theta})|^2 \, d\theta \leqslant M^2;$$

and since this holds for all $r < 1$ our assertion is proved.

Now by the Riesz–Fischer theorem

$$\sum_{n=0}^{\infty} |a_n|^2 = O(1)$$

implies that the series $\sum_{n=0}^{\infty} a_n e^{in\theta}$

is the Fourier series of a function $f(\theta)$ (Titchmarsh [1], p. 423) and is therefore summable $(C, 1)$ to $f(\theta)$ for almost all θ in $0 \leqslant \theta \leqslant 2\pi$. But a series which is summable $(C, 1)$ is Abel summable (Hardy [1], p. 108, theorem 55). That is to say

$$\lim_{r \to 1} f(r e^{i\theta}) = \lim_{r \to 1} \sum_{n=0}^{\infty} a_n r^n e^{in\theta}$$

exists for almost all θ in $0 \leqslant \theta \leqslant 2\pi$, and this proves the theorem.

COROLLARY 1. *If the function $u(z) = u(r, \theta)$ is harmonic and bounded in $|z| < 1$, then the radial limits $u(e^{i\theta}) = \lim_{r \to 1} u(r, e^{i\theta})$ exist for all points $e^{i\theta}$ on $|z| = 1$, except possibly for a set of linear measure zero.*

For $u(z)$ is the real part of a function $f(z)$ which is regular in $|z| < 1$. For a finite real number $A > u(z)$ for all $|z| < 1$ the function $(f(z) - A)^{-1}$ is regular and bounded and therefore has radial limits at almost all points $e^{i\theta}$ on $|z| = 1$. The same is therefore true of $f(z)$ and of $u(z) = \mathrm{Re} f(z)$.

3. Lindelöf's theorem and Fatou's theorem for angular limits

We remark that Fatou's theorem, as we have stated it, guarantees the existence of radial limit values of $f(z)$ almost everywhere on $|z| = 1$. This theorem can be generalized in many ways, by relaxing the condition that $|f(z)|$ be bounded in $|z| < 1$, for example. We shall indicate some of these possibilities, but, at this stage, we wish to show that, for a function analytic and bounded in $|z| < 1$, the existence of a radial limit at a point $e^{i\theta}$ implies the existence of the limit for any non-tangential approach to $e^{i\theta}$.

To this end we state next a theorem of Lindelöf [1], together with two immediate corollaries, which will enable us to state Theorem 2.1 in a form (Theorem 2.4) having considerable application in the sequel. This result of Lindelöf has many applications in the theory of functions.

THEOREM 2.2. *Let $f(z)$ be analytic and bounded in an angular domain T: $|\arg z| < \theta_0$, $0 < |z| < R_0$, and let $f(z)$ tend to a unique value α as $z \to 0$ along a continuous path \mathscr{L} lying in T and terminating at $z = 0$. Then $f(z)$ tends to α uniformly as $z \to 0$ inside any angular subdomain of opening $-\theta_0 + \delta \leqslant \arg z \leqslant \theta_0 - \delta$.*

By an elementary conformal mapping, we may state Theorem 2.2 in the following form:

THEOREM 2.3. *Let $f(z)$ be analytic and bounded in $|z| < 1$. If $f(z) \to \alpha$ as $z \to e^{i\theta_0}$ along some arc \mathscr{L} lying in $|z| < 1$ and terminating at $e^{i\theta_0}$, then $f(z)$ tends to α uniformly as $z \to e^{i\theta_0}$ inside any angular domain lying in $|z| < 1$ and having $e^{i\theta_0}$ as vertex.*

The theorem of Lindelöf (Theorem 2.3) may be easily generalized in various directions; for example, we may replace the hypothesis of boundedness by the hypothesis that $\mathrm{Re} f(z) < c$, for some

finite c. We prove here an extension† of Lindelöf's theorem involving two paths:

THEOREM 2.3.1. *Let $f(z) = u(z) + iv(z)$ be analytic and bounded in $|z| < 1$, and let \mathscr{L}_1 and \mathscr{L}_2 be two arcs lying in $|z| < 1$ and terminating at a point P of $|z| = 1$. If $u(z) \to a$ as $z \to P$ on \mathscr{L}_1 and $v(z) \to b$ as $z \to P$ on \mathscr{L}_2, then $f(z)$ tends to $a + ib$ uniformly as $z \to P$ inside any angular domain of opening $\pi - \epsilon$ having vertex at P and bisected by the radius drawn to P.*

By considering the function $g(z) = f(z) - a - ib$, we may assume that the real [imaginary] part of $g(z)$ tends to 0 as $z \to P$ along \mathscr{L}_1 [\mathscr{L}_2]; we may also assume that P is the point $z = 1$. Suppose first that $\mathscr{L}_1 \cap \mathscr{L}_2$ is empty in a neighbourhood of $z = 1$. Then, by a conformal mapping, the theorem can be reduced to the following. Suppose that $g(z)$ is analytic and bounded on $U: |z| < 1$, and that the Fatou boundary function $\operatorname{Re} g(e^{i\theta})$ [$\operatorname{Im} g(e^{i\theta})$] of the real [imaginary] part of g has limit 0 at $z = 1$ on approach from below [above]. It is to be proved that g has radial limit 0 at $z = 1$. Now the function $g_1(z) = \bar{g}(\bar{z})$ is analytic and bounded in U, so that gg_1 is analytic and bounded in U; furthermore, the real part of the Fatou boundary function has limit 0 at $z = 1$, which is a point of continuity. Therefore, by a well-known theorem of Fejér (see, for example, Titchmarsh [1], p. 414), the Fourier series for gg_1 is summable $(C, 1)$ at $z = 1$ and its real part has radial limit 0 there; that is, $|g|^2$ has radial limit 0 at $z = 1$, as was to be proved.

If \mathscr{L}_1 and \mathscr{L}_2 intersect arbitrarily close to $z = 1$, we proceed as follows, assuming, as we may, that each of \mathscr{L}_1 and \mathscr{L}_2 is polygonal and does not intersect itself. These paths determine simple closed polygons. Let \mathscr{L}_i' be that part of \mathscr{L}_i, $i = 1, 2$, on one such polygon, and let z_1, z_2 be the end-points of \mathscr{L}_i. Let $|u - a|$ [$|v - b|$] be bounded by c on \mathscr{L}_1' [\mathscr{L}_2'], and let M be the supremum of $|g|$ in U. The theorem will be proved by an application of Theorem 2.3 if we can exhibit a continuous path \mathscr{L}' in the polygon, from z_1 to z_2, on which $|g|$ is small with c, and, in fact, we shall obtain

$$|g|^2 \leqslant 2cM \tag{2.3.1}$$

† Gehring and Lohwater [1]. The present proof is due to Doob and avoids the deep theorem of Gross [4] on which the original proof was based. For a more recent and transparent proof of Gross's theorem, see Doob [13].

on \mathscr{L}'. We map conformally the interior of the polygon onto U in such a way that z_1 and z_2 go into ± 1, and we define \mathscr{L}' to be the inverse image of the diameter through ± 1. An argument based on the maximum principle for harmonic functions shows that the transform of g satisfies (2.3.1) on the diameter through ± 1. Hence g satisfies (2.3.1) on \mathscr{L}', as was to be proved.

To prove this we observe, writing $f(\xi)$, $|\xi| \leqslant 1$, for the transform of $g(z)$ and $f_1 = \bar{f}(\bar{\xi})$, that $\mathrm{Re} f < c$ on $|z| = 1$, $\mathrm{Im}\,\xi \leqslant 0$; $\mathrm{Im} f < c$ on $|\xi| = 1$, $\mathrm{Im}\,\xi \geqslant 0$; $|f| < M$ on $|\xi| \leqslant 1$. Hence

$$\mathrm{Re}\, ff_1 < 2cM \quad \text{on} \quad |\xi| = 1. \tag{2.3.2}$$

Now ff_1 is analytic so that $\mathrm{Re}\, ff_1$ is harmonic on the closed disc and hence satisfies (2.3.2) on the diameter through ± 1. But on this diameter
$$ff_1 = |f|^2 = \mathrm{Re}\, ff_1 < 2cM,$$

which proves our assertion.

By combining Theorems 2.1 and 2.3, we obtain the following stronger form of Fatou's theorem, where the radial limits are replaced by angular limits.

THEOREM 2.4. *Let $f(z)$ be analytic and bounded in $|z| < 1$. Then, for all $e^{i\theta}$ on $|z| = 1$, except for a possible set of linear measure zero, and for any $\epsilon > 0$, $f(z)$ tends uniformly to a limit as $z \to e^{i\theta}$ inside an angular domain of opening $\pi - \epsilon$ having vertex at $e^{i\theta}$ and bisected by the radius drawn to $e^{i\theta}$.*

We shall call a point $e^{i\theta}$ a *Fatou point* of a function $f(z)$, meromorphic in $|z| < 1$, if $f(z)$ tends uniformly to some limit (which may be ∞) inside every angular domain of opening $\pi - \epsilon$ having vertex at $e^{i\theta}$ and bisected by the radius drawn to $e^{i\theta}$. The set of all Fatou points of $f(z)$ is denoted by $F(f)$ and the corresponding set $\bigcup_{e^{i\theta} \in F(f)} f(e^{i\theta})$, where $f(e^{i\theta})$ is the angular limit of $f(z)$ at $e^{i\theta}$, by $\Gamma_F(f)$.

4. Theorem of F. and M. Riesz

We have seen in § 3 that the radial limits of a bounded analytic function exist for almost all points of $|z| = 1$; we shall state in this section a uniqueness theorem of F. and M. Riesz [1] from which it follows that the radial limit values cannot form too sparse a set

in the w-plane. The theorem is a special case of Theorem 2.19, a proof of which we shall give in outline in § 9 (p. 41). We have

THEOREM 2.5. *If $f(z)$ is analytic and bounded, $|f(z)| < M$, in $|z| < 1$, and if the set E of points $e^{i\theta}$ for which $f(e^{i\theta}) = \lim\limits_{r \to 1} f(r\,e^{i\theta}) = 0$ has positive measure on $|z| = 1$, then $f(z)$ is identically zero in $|z| < 1$.*

We remark that an immediate consequence of Theorem 2.5 is that two bounded analytic functions having the same radial limit values on a set of positive measure must be identical.

As an immediate corollary to Theorem 2.5 we have,

COROLLARY 2. *For a non-constant function $f(z)$, analytic and bounded in $|z| < 1$, the set $\Gamma_F(f)$ is non-countable.*

5. Existence theorems on radial limits

The theorem of Fatou is capable of extension in several directions; for example, we shall show later in this chapter that the restriction of boundedness may be replaced by requiring the function merely to be the quotient of two bounded analytic functions, while in Chapter 3 we shall show that the set of points on $|z| = 1$ at which $f(z)$ does *not* possess radial limits is substantially more sparse, i.e. of capacity zero, if $f(z)$ is restricted to be univalent or schlicht in $|z| < 1$.

Before proceeding to discuss other forms of Fatou's theorem, however, we wish to show that, without further conditions on $f(z)$, Theorem 2.1 is best possible in the sense that the size of the exceptional set cannot be further reduced. It was first shown in 1919 by N. Lusin [1]† that for every set E of measure zero on $|z| = 1$ there exists a function $f(z)$, analytic and bounded in $|z| < 1$, such that $f(z)$ has no radial limit along a radius terminating in a point of E. This result of Lusin is connected intimately with a more general problem which we shall discuss in this section:

> *What are the necessary and sufficient conditions to be imposed upon a set E on $|z| = 1$, in order that there exist a function $f(z)$, analytic and bounded in $|z| < 1$, such that the radial limits of $f(z)$ exist for each point of $\mathscr{C}E$ and for no point of E?*

† The example which appears in the more widely known paper of Lusin and Privalov [1] is essentially the same.

From Theorem 2.1 it follows, of course, that E must be of linear measure zero, and the following argument shows that an additional restriction of a topological character must also be imposed.

We say that a set E is *of type F_σ* if E is the union of a countable set of closed sets, and that a set E is *of type G_δ* if E is the intersection of countably many open sets. A set E which is the intersection of a countable set of sets of type F_σ will be called of type $F_{\sigma\delta}$, while the union of a countable set of sets of type G_δ will be called of type $G_{\delta\sigma}$. With this concept, it is easy to establish the following result.

THEOREM 2.6. *If $f(z) = f(r\,e^{i\theta})$ is continuous in $|z| < 1$, then the set E of points $e^{i\theta}$ on $|z| = 1$ where the radial limit $\lim\limits_{r\to 1} f(r\,e^{i\theta})$ fails to exist is of type $G_{\delta\sigma}$.*

Indeed, for every $\epsilon > 0$, there is a δ, $0 < \delta < 1$, such that for $e^{i\theta}$ not in E
$$|f(r_1 e^{i\theta}) - f(r_2 e^{i\theta})| \leqslant \epsilon \quad \text{for} \quad r_1 > r_2 > \delta. \qquad (2.5.1)$$
For fixed ϵ and δ, the set $F(\epsilon, \delta)$ of points $e^{i\theta}$ for which (2.5.1) holds is closed, being the intersection of those sets for which (2.5.1) holds for a fixed pair r_1 and r_2. Now the set of $e^{i\theta}$ for which (2.5.1) holds for fixed ϵ and arbitrary δ is of type F_σ, since we may write it as
$$F(\epsilon) = \bigcup_\delta F(\epsilon, \delta) = \bigcup_{n=1}^\infty F(\epsilon, r_n),$$
where $\{r_n\}$ is a monotone increasing sequence such that $\lim\limits_{n\to\infty} r_n = 1$. Finally, the set of $e^{i\theta}$ for which the radial limit $\lim\limits_{r\to 1} f(r\,e^{i\theta})$ exists can be written as the intersection
$$\bigcap_{\epsilon>0} F(\epsilon) = \bigcap_{n=1}^\infty F(\epsilon_n),$$
where $\{\epsilon_n\}$ is a monotone decreasing sequence such that $\lim\limits_{n\to\infty} \epsilon_n = 0$. Thus the set of points $e^{i\theta}$ for which $\lim\limits_{r\to 1} f(r e^{i\theta})$ exists is of type $F_{\sigma\delta}$, so that its complement E on $|z| = 1$ is of type $G_{\delta\sigma}$.

What the example of Lusin shows, then, in light of Theorem 2.6, is that every set E of measure zero on $|z| = 1$ is contained in a set E_1 of type $G_{\delta\sigma}$, also of measure zero, with the property that E_1 is *precisely* the set on which some bounded analytic function

fails to possess radial limits. We prove next a theorem which generalizes the construction of Lusin, and which is a step nearer the solution of this general problem; we must remark, however, that the general problem has not been solved completely and that the following theorem† constitutes the best result known today.

THEOREM 2.7. *Let the set E on $|z| = 1$ be of type F_σ and of linear measure zero. Then there exists a function $f(z)$, analytic and bounded in $|z| < 1$, such that the radial limit $f(e^{i\theta})$ exists for each $e^{i\theta}$ in $\mathscr{C}E$ and for no $e^{i\theta}$ in E.*

To prove Theorem 2.7, we require two lemmas, the first of which sharpens the result of Lusin which we have been discussing.

LEMMA 1. *Let E be a set on $|z| = 1$ of linear measure zero. Then there exists a function $f(z)$, analytic and bounded, $|f(z)| < 1$, whose radial limits $f(e^{i\theta})$ have modulus 1 for almost all points of $|z| = 1$, and such that*

$$\liminf_{r \to 1} |f(r\,e^{i\theta})| = 0, \quad \limsup_{r \to 1} |f(r\,e^{i\theta})| = 1, \qquad (2.5.2)$$

for each $e^{i\theta}$ in E.

REMARK. Condition (2.5.2) may be written in the form

$$C_\rho(|f|, e^{i\theta}) = [0, 1] = C(|f|, e^{i\theta}),$$

where \mathscr{G} in Theorem 1.1 is taken to be the radius ρ to $e^{i\theta}$. For, since $f(z)$ is continuous in $|z| < 1$, the whole segment $[0, 1]$ is the radial cluster set and since $|f| < 1$ it must also be the complete cluster set at $e^{i\theta}$. $C_\rho(|f|, e^{i\theta})$ is thus *maximal* at all points of E. We shall be discussing this idea of a maximal partial cluster set more fully in Chapter 4.

The idea behind the proof of Lemma 1 is quite simple. We construct a function $g_1(z)$, whose real part is positive in $|z| < 1$ and tends to zero on almost all radii and to infinity on a set of special radii. These special radii terminate 'near' the set E, but not in E. The special radii cause the real part of $g_1(z)$ to assume, on those radii terminating in E, a fairly large value and then to recede to zero. Then, relative to a second set of special radii, a second function $g_2(z)$ is constructed with a similar behaviour, with the difference that, on those radii terminating in E, the real part of $g_2(z)$

† Lohwater and Piranian [1].

remains small until the real part of $g_1(z)$ has reached its peak and has become small. The construction continues in this fashion, and the function

$$f(z) = \exp\{-\Sigma g_n(z)\}$$

has the desired properties. The details now follow.

Let $\{\epsilon_n\}$ be a sequence of positive numbers with the properties $\frac{1}{3} > \epsilon_1 > \epsilon_2 > ... > 0$, $\Sigma \epsilon_n < 1$; we note that these properties imply that $n\epsilon_n \to 0$. Let G_1 be an open set on $|z| = 1$ containing E such that $m(G_1) < \epsilon_1^2$, where $m(G_1)$ denotes the linear measure of G_1, and let $\{I_{1k}\}$ denote the family of disjoint open arcs forming G_1. For each k, let z_{1k} denote one of the end-points of I_{1k}, let us say the left end-point as viewed from the origin. We define the function $g_1(z)$ by the formula

$$g_1(z) = \frac{1}{\epsilon_1} \sum_k m(I_{1k}) \frac{z_{1k}+z}{z_{1k}-z}. \tag{2.5.3}$$

On the circle C_{1k} of diameter $2m(I_{1k})$ and internally tangent to the circumference $C: |z| = 1$ at z_{1k}, the real part of the kth term under the summation sign has the constant value

$$m(I_{1k})(2-2m(I_{1k}))/2m(I_{1k}) = 1 - m(I_{1k}) > \tfrac{1}{2}.$$

Since the real part of each term of the sum (2.5.3) is positive in $|z| < 1$, the real part of $g_1(z)$ is greater than $1/2\epsilon_1$ on C_{1k}.

Let C_{1k}^* denote the circle of diameter $2 - \epsilon_1$ and internally tangent to C at z_{1k}, and let R_1 denote the set of all points lying outside all the circles C_{1k}^*. Let C_1 denote the set of those boundary points of R_1 lying in $|z| < 1$. On C_1, the kth term of the sum (2.5.3) has real part less than $m(I_{1k})\epsilon_1/2 - \epsilon_1$; consequently, the real part of $g_1(z)$ is less than

$$(2-\epsilon_1)^{-1}\Sigma m(I_{1k}) < \frac{\epsilon_1^2}{2-\epsilon_1} < \epsilon_1(1-\epsilon_1)$$

at each point of C_1.

Suppose next that we have defined, for each of the indices $m = 1, 2, ..., n-1$, the open set G_m, the finite or countable set $\{z_{mk}\}$, the region R_m with boundary C_m relative to $|z| < 1$, and the function $g_m(z)$ such that on the set C_m the inequality

$$\operatorname{Re} \sum_{k=1}^{n-1} g_k(z) < m.\epsilon_m(1-\epsilon_{m-1}) \quad (m = 1, 2, ..., n-1)$$

holds, and such that no point of G_m lies in the closure of C_m. We shall now construct an open set G_n, $G_{n-1} \supset G_n \supset E$, $m(G_n) < \epsilon_n^2$, and the corresponding function

$$g_n(z) = \frac{1}{\epsilon_n} \sum_k m(I_{nk}) \frac{z_{nk}+z}{z_{nk}-z}, \qquad (2.5.4)$$

where the meanings of the symbols I_{nk} and z_{nk} are analogous to those of I_{1k} and z_{1k}. In order that the behaviour of the function $g_n(z)$ does not interfere appreciably with that of the function $\sum_1^{n-1} g_m(z)$ on the sets $C_1, C_2, ..., C_{n-1}$, it will be necessary to make the set G_n sufficiently thin in an appropriate sense.

Let each arc $I_{n-1,h}$ be divided into subarcs $I_{n-1,h,j}$ $(j = 1, 2, ...)$ having both end-points in E, and such that the distance $d_{n-1,h,j}$ between $I_{n-1,h,j}$ and $C_1 \cup C_2 \cup ... \cup C_{n-1}$ is positive. Since $m(E) = 0$, we may subject G_n to the restriction

$$m(G_n \cap I_{n-1,h,j}) < \frac{d_{n-1,h,j}(\epsilon_{n-1}-\epsilon_n)\epsilon_n^2}{6h^2j^2}.$$

Now for each point in C_m $(m = 1, 2, ..., n-1)$ and each pair j, h, the sum of moduli of those terms of (2.5.4) arising from points z_{nk} in $I_{n-1,h,j}$ is less than

$$\frac{2m(G_n \cap I_{n-1,h,j})}{d_{n-1,h,j}} < \frac{\epsilon_n^2(\epsilon_{n-1}-\epsilon_n)}{3h^2j^2}.$$

Since
$$\sum_{n,j} \frac{1}{3h^2j^2} < 1,$$

it follows that $|g_n(z)| < \epsilon_n(\epsilon_{n-1}-\epsilon_n)$

on C_m, so that, at points of C_m $(m = 1, 2, ..., n)$,

$$\text{Re} \sum_{k=1}^n g_k(z) < m \cdot \epsilon_m(1-\epsilon_{n-1}) + \epsilon_n(\epsilon_{n-1}-\epsilon_n) \leqslant m \cdot \epsilon_m(1-\epsilon_n).$$

The remainder of the proof of the lemma goes rather easily. On each circle C_{nk}, the real part of $g_n(z)$ is greater than $1/2\epsilon_n$. Let R_n be the region exterior to each of the circles which have diameter $2 - \epsilon_n$ and are internally tangent to C at one of the points z_{mk} $(m = 1, 2, ..., n; \ k = 1, 2, ...)$, and let C_n consist of those

boundary points of R_n which lie in $|z| < 1$. At each point of C_n, the real part of each of the quantities $g_m(z)$ $(m = 1, 2, ..., n)$ is less than

$$\frac{\epsilon_m \epsilon_n}{2 - \epsilon_n} < \epsilon_n(1 - \epsilon_n),$$

so that

$$\text{Re} \sum_{m=1}^{n} g_m(z) < n \cdot \epsilon_n(1 - \epsilon_n).$$

Furthermore, each point of G_n is at a positive distance from the sets $C_1, C_2, ..., C_n$ so that the construction can be continued.

It is clear that the series $\Sigma g_n(z) = g(z)$ converges uniformly on every closed subset of $|z| < 1$ and thus represents an analytic function in $|z| < 1$. Since $\text{Re}\, g(z)$ is positive in $|z| < 1$, the function $f(z) = e^{-g(z)}$ is analytic and bounded in $|z| < 1$. Every radius of $|z| < 1$ terminating in E must pass through each of the sets C_n, and, between each pair of intersections with two of the sets C_{n-1} and C_n, it must pass through one of the circles C_{nk}, so that the condition (2.5.2) is satisfied on every such radius.

Finally, we remark that, because $g(z)$ is of the form

$$g(z) = \sum_k b_k \frac{e^{i\theta_k} + z}{e^{i\theta_k} - z},$$

where b_k is a positive constant such that $\sum_k b_k < \infty$, it follows that $\lim_{r \to 1} \text{Re}\, g(r\, e^{i\theta}) = 0$ for almost all $e^{i\theta}$ on $|z| = 1$, so that $\lim_{r \to 1} |f(r\, e^{i\theta})| = 1$ almost everywhere. This completes the proof of Lemma 1.

LEMMA 2. *If E is a closed set of measure zero on $|z| = 1$, there exists a function $f(z)$, analytic and bounded in $|z| < 1$, $|f(z)| < 1$, such that the radial limit $f(e^{i\theta})$ exists and has modulus 1 or 0 for every $e^{i\theta}$ in $\mathscr{C}E$, and such that condition (2.5.2) is satisfied for each $e^{i\theta}$ in E.*

The remark after Lemma 1 applies also to Lemma 2.

Since the set E is closed, the construction in the proof of Lemma 1 may be repeated with the additional requirement that each arc I_{nk} contain a point of E and that, for each index n, the set $\{z_{nk}\}$ be finite. If $e^{i\theta}$ lies in $\mathscr{C}E$, it is either one of the points z_{nk}, or it lies at a positive distance from the set $\{z_{nk}\}$. In the first case,

$f(e^{i\theta}) = 0$. In the second case, the series $\Sigma g_n(z)$ converges uniformly on the radius drawn to $e^{i\theta}$, and, since the real part of $g_n(z)$ has radial limit zero on this radius, the same is true of the sum; therefore, $f(e^{i\theta})$ exists and has modulus 1. This concludes the proof of Lemma 2.†

To conclude the proof of Theorem 2.7 we make use of the fact that the function $f(z)$ of Lemma 2 has a uniformly large oscillation on the radii terminating in the pre-assigned closed set. We suppose now that $E = \bigcup_n F_n$, where the sets F_n are closed and of measure zero. For each n, we denote by $f_n(z)$ a function having the properties described in Lemma 2, relative to the set F_n. The function $f(z) = \sum_n 3^{-n} f_n(z)$ has the properties asserted in Theorem 2.7.

Blaschke products

6. Product representation of a bounded function

We consider in the next two sections an important class of functions which are analytic and bounded in $|z| < 1$; although one is led to the consideration of this class in many ways, we prefer to approach the notion by means of the properties of the zeros of bounded analytic functions. For our discussion of the properties of the zeros of a bounded function, we shall assume that $|f(z)| < 1$ in $|z| < 1$, for otherwise we consider $f(z)/M$, where $M = \sup |f(z)|$. We shall assume in the first part of our discussion that $f(0) \neq 0$; otherwise, we treat the function $f(z)/z^\lambda$. Let $\{a_k\}$, $k = 1, 2, \ldots$, denote the zeros of $f(z)$, which we assume are so arranged that $0 < |a_k| \leqslant |a_{k+1}| < 1$. The function

$$b_n(z) = \prod_{k=1}^{n} |a_k| \frac{1 - (z/a_k)}{1 - \bar{a}_k z} \qquad (2.6.1)$$

has the property that it is analytic in $|z| \leqslant 1$ with modulus 1 on $|z| = 1$. Thus the function $f(z)/b_n(z)$ is analytic in $|z| < 1$ with $|f(z)/b_n(z)| \leqslant 1$. Consequently, we have $|f(0)| \leqslant |b_n(0)| < 1$, or

$$0 < |f(0)| \leqslant \prod_{k=1}^{n} |a_k| < 1. \qquad (2.6.2)$$

Now (2.6.2) implies the convergence of the infinite product

† For an interesting existence theorem related to this lemma see Ohtsuka, *Proc. Amer. Math. Soc.* 5 (1954), 533–535.

$\prod\limits_{k=1}^{\infty} |a_k|$, and, since the convergence of this infinite product is equivalent to the convergence of the sum $\sum\limits_{k=1}^{\infty} (1 - |a_k|)$, we have

THEOREM 2.8. *Let $f(z)$ be analytic and bounded in $|z| < 1$, and let $\{a_k\}$, $k = 1, 2, \ldots$ be the zeros of $f(z)$. Then the series*

$$\sum_{k=1}^{\infty} (1 - |a_k|) \tag{2.6.3}$$

converges.

By way of a converse to Theorem 2.8, we shall prove that, if the sequence $\{a_k\}$ is such that the series (2.6.3) converges, there exists a bounded analytic function in $|z| < 1$ with $\{a_k\}$ as its zeros. It is clear that the convergence of the infinite product

$$B_0(z) = \lim_{n \to \infty} \prod_{k=1}^{n} |a_k| \frac{1 - (z/a_k)}{1 - \bar{a}_k z} \tag{2.6.4}$$

depends on the convergence of the infinite series

$$\sum_{k=1}^{\infty} \left\{ |a_k| \frac{1 - (z/a_k)}{1 - \bar{a}_k z} - 1 \right\} = \sum_{k=1}^{\infty} \frac{(|a_k| - 1)(1 + (|a_k|/a_k)z)}{1 - \bar{a}_k z}. \tag{2.6.5}$$

Thus, for $|z| \leqslant r_0 < 1$, the modulus of the general term of (2.6.5) is less than

$$(1 - |a_k|) \frac{1 + r_0}{1 - r_0},$$

so that (2.6.5) converges uniformly and absolutely in $|z| \leqslant r_0$. This implies that the product in (2.6.4) converges to a function $B_0(z)$ which is analytic and bounded, $|B_0(z)| < 1$, in $|z| < 1$, which we shall call a *Blaschke product* with zeros $\{a_k\}$. We remark that the restriction that $a_k \neq 0$ is only temporary, and can be discarded by considering the function $z^\lambda B_0(z) = B(z)$.

We consider again the analytic, bounded function $f(z)$ having zeros at the points $\{a_k\}$, $k = 1, 2, \ldots$; $a_k \neq 0$. If $|f(z)| < 1$ in $|z| < 1$, and if $B_0(z)$ is a Blaschke product with zeros $\{a_k\}$, then the function $f(z)/B_0(z)$ is analytic without zeros in $|z| < 1$, and, in addition,

$$\left| \frac{f(z)}{B_0(z)} \right| = \lim_{n \to \infty} \left| \frac{f(z)}{b_n(z)} \right| \leqslant 1$$

in $|z| < 1$. Thus we may write

$$f(z) = B_0(z)\, e^{g(z)},$$

where $g(z)$ is analytic with negative real part in $|z| < 1$. Finally, we remove the restriction that $f(0) \neq 0$ by remarking that, if $f(z)$ has a zero of order λ at $z = 0$, the discussion above is applicable to the function $f(z)/z^\lambda$, so that we have

THEOREM 2.9. *Every function which is analytic and bounded,* $|f(z)| < 1$, *in* $|z| < 1$ *has a representation of the form*

$$f(z) = e^{g(z)} z^\lambda \prod_{k=1}^{\infty} |a_k| \frac{1 - (z/a_k)}{1 - \bar{a}_k z} = B(z)\, e^{g(z)}, \qquad (2.6.6)$$

where $g(z)$ *is analytic with negative real part in* $|z| < 1$.

The function $F(z) = 1/[1 - g(z)]$ is then analytic and bounded in $|z| < 1$, so that, by Theorem 2.1, $F(z)$ has radial limits almost everywhere; by Theorem 2.5, $g(z)$ must possess finite radial limits almost everywhere. If $u(r, \theta)$ denotes the real part of $g(z)$, the function

$$\mu(r, \theta) = \int_0^\theta u(r, t)\, dt$$

is a monotone non-increasing function of θ for each fixed r, and

$$\int_0^{2\pi} |u(r, t)|\, dt = -\int_0^{2\pi} u(r, t)\, dt = -2\pi u(0, 0)$$

is uniformly bounded in r. The limit

$$\mu(\theta) = \lim_{r \to 1} \int_0^\theta u(r, t)\, dt$$

exists except for a possibly countable set of values of θ, and defines a monotone non-increasing function of θ. It is a classical extension† of Poisson's theorem, which we assume, that $u(r, \theta)$ has an integral representation of the form

$$u(r, \theta) = \frac{1}{2\pi} \int_0^{2\pi} \frac{1 - r^2}{1 + r^2 - 2r \cos(\theta - t)}\, d\mu(t), \qquad (2.6.7)$$

where $\mu(\theta)$ is monotone non-increasing in $0 \leqslant \theta \leqslant 2\pi$, and with

† Golusin [1], pp. 333–345, contains a detailed discussion which we shall not go into here.

the property that $\lim\limits_{r \to 1} u(r, \theta) = \mu'(\theta)$ at any point where $\mu'(\theta)$ exists. Thus we have

THEOREM 2.10. *Every function $f(z)$, analytic and bounded, $|f(z)| < 1$, in $|z| < 1$ has a Poisson–Stieltjes integral representation,*

$$f(z) = e^{i\alpha}z^\lambda \prod_{k=1}^{\infty} |a_k| \frac{1 - (z/a_k)}{1 - \bar{a}_k z} \exp\left\{\frac{1}{2\pi}\int_0^{2\pi} \frac{e^{it} + z}{e^{it} - z} d\mu(t)\right\}, \quad (2.6.8)$$

where $\mu(\theta)$ is monotone non-increasing in $0 \leqslant \theta \leqslant 2\pi$, where α is a constant, and where the non-vanishing zeros $\{a_k\}$ of $f(z)$ satisfy

$$\sum_{k=1}^{\infty} (1 - |a_k|) < \infty.$$

7. Boundary properties of Blaschke products

The class of functions known as Blaschke products is distinguished by many interesting features, some of which we develop in this section. The first of these properties, due to F. Riesz [1], is that the radial limit values, which exist almost everywhere by Fatou's theorem, are of modulus 1 almost everywhere on $|z| = 1$. Now for a Blaschke product $B(z)$, we have $|B(z)| < 1$ in $|z| < 1$, and, since the coefficient of z^λ in the Taylor series expansion $B(z) = C_\lambda z^\lambda + \dots$ is $\prod\limits_{k=1}^{\infty} |a_k|$, the Poisson–Jensen formula,† when applied to $B(z)$, yields

$$\frac{1}{2\pi}\int_0^{2\pi} \log |B(r\,e^{i\theta})| \, d\theta = \log |C_\lambda| + \lambda \log r + \log \prod_{0 < |a_k| < r} \frac{r}{|a_k|}$$

$$= (\lambda + n(r)) \log r + \log \prod_{|a_k| \geqslant r} |a_k|,$$

where $n(r)$ is the number of points a_k lying inside $|z| < r$. The convergence of (2.6.3) implies that $\lim\limits_{n \to \infty} n(1 - |a_n|) = 0$, or equivalently, that $\lim\limits_{n \to \infty} n \log |a_n| = 0$, so that

$$\lim_{r \to 1} n(r) . \log r = 0. \quad (2.7.1)$$

Hence we have

$$\lim_{r \to 1} \int_0^{2\pi} \log |B(r\,e^{i\theta})| \, d\theta = 0. \quad (2.7.2)$$

† Titchmarsh [1], p. 129.

Now Fatou's theorem, together with the inequality $|B(z)| < 1$, implies that $\log|B(e^{i\theta})| \leqslant 0$ almost everywhere; we show that $\log|B(e^{i\theta})| = 0$ almost everywhere. Let us suppose, to the contrary, that $\log|B(e^{i\theta})| < 0$ on a set of positive measure on $|z| = 1$. Then there exists an integer n_0 such that $\log|B(e^{i\theta})| < -1/n_0$ on a set E of positive measure. By Egorov's theorem, there exists a closed subset E_0 of E of positive measure, such that the convergence of $B(r\,e^{i\theta})$ to $B(e^{i\theta})$ is uniform† on E_0, and such that $\log|B(r\,e^{i\theta})| < -1/2n_0$ for $r > r_0$ and $e^{i\theta}$ in E_0. Thus, for $r_0 < r < 1$, we obtain the inequality

$$\int_0^{2\pi} \log|B(r\,e^{i\theta})|\,d\theta \leqslant \int_{E_0} \log|B(r\,e^{i\theta})|\,d\theta < \frac{-m(E_0)}{2n_0}.$$

When r approaches 1, this inequality implies that

$$\lim_{r\to 1} \int_0^{2\pi} \log|B(r\,e^{i\theta})|\,d\theta = 0 < \frac{-m(E_0)}{2n_0},$$

which is absurd, so that $\log|B(e^{i\theta})|$ cannot be negative on a set of positive measure. Thus we have

THEOREM 2.11. *A Blaschke product possesses radial limits of modulus 1 for almost all $e^{i\theta}$ on $|z| = 1$.*

By means of Theorem 2.11 we shall be able to give a necessary and sufficient condition that a bounded function be a Blaschke product.

THEOREM 2.12. *A necessary and sufficient condition that a function $f(z)$, analytic and bounded in $|z| < 1$, $|f(z)| < 1$, be a Blaschke product is that*

$$\lim_{r\to 1} \int_0^{2\pi} \log|f(r\,e^{i\theta})|\,d\theta = 0. \qquad (2.7.3)$$

It is clear that, if $f(z)$ is a Blaschke product, then (2.7.3) is nothing more than (2.7.2).

We show, conversely, that if (2.7.3) holds, then $f(z)$ is a Blaschke product. Now (2.6.6) implies that

$$\log|f(r\,e^{i\theta})| = u(r,\theta) + \log|B(r\,e^{i\theta})|,$$

† The transition from the sequential formulation of Egorov's theorem to the present situation is easily made.

where $u(r, \theta) = \operatorname{Re} g(z)$, whence

$$
\begin{aligned}
0 &= \lim_{r \to 1} \int_0^{2\pi} \log \left| f(r e^{i\theta}) \right| d\theta \\
&= \lim_{r \to 1} \int_0^{2\pi} u(r, \theta)\, d\theta + \lim_{r \to 1} \int_0^{2\pi} \log \left| B(r e^{i\theta}) \right| d\theta = 2\pi \cdot u(0,0).
\end{aligned}
$$

Since $u(r, \theta)$ is non-positive in $|z| < 1$, the condition $u(0,0) = 0$ implies that $u(r, \theta)$ is identically zero, so that, by (2.6.6), $f(z)$ reduces to a Blaschke product.

We remark first that the condition of boundedness may be dropped from Theorem 2.12 if the condition (2.7.3) be replaced by

$$
\lim_{r \to 1} \int_0^{2\pi} \left| \log \left| f(r e^{i\theta}) \right| \right| d\theta = 0. \tag{2.7.4}
$$

We remark further that the existence of radial limits of modulus 1 almost everywhere does not distinguish the class of Blaschke products in the class of bounded analytic functions, for the function $f(z) = e^{(z+1)/(z-1)}$ is analytic and bounded, $|f(z)| < 1$, in $|z| < 1$ and possesses a radial limit of modulus 1 at each point of $|z| = 1$, except at $z = 1$, where 0 is a radial limit. Indeed, one of the powerful methods in the sequel will consist of an analysis of those bounded analytic functions which have radial limits of modulus 1 almost everywhere and which are not Blaschke products. Before proceeding to an analysis of such functions, we complete our discussion of the radial limits of Blaschke products. From the following theorem† we see that a Blaschke product $B(z)$ can possess a radial limit of modulus 1 at *every* point of $|z| = 1$ even though $B(z)$ has infinitely many zeros.

THEOREM 2.13. *Let $B(z)$ be a Blaschke product whose zeros $\{a_k\}$ have $e^{i\theta_0}$ as a limit point. Then, a necessary and sufficient condition that $B(z)$ and all its partial products have radial limits of modulus 1 at $e^{i\theta_0}$ is the convergence of*

$$
\sum_{k=1}^{\infty} \frac{d_k}{\left| e^{i\theta_0} - a_k \right|}, \tag{2.7.5}
$$

where $d_k = 1 - |a_k|$.

† Frostman [2].

We lose no generality by assuming that $e^{i\theta_0} = 1$ and that $z = r < 1$. Since

$$|1 - \bar{a}_k r| > 1 - r \quad \text{and} \quad |1 - \bar{a}_k r| > \tfrac{1}{2}|1 - a_k|,$$

the series
$$\sum_{k=1}^{\infty} \frac{(1 - r^2)(1 - |a_k|^2)}{|1 - \bar{a}_k r|^2}$$

converges uniformly, in view of (2.7.5). Thus, in the inequality

$$|B(r)|^2 = \prod_{k=1}^{\infty} \left\{ 1 - \frac{(1 - r^2)(1 - |a_k|^2)}{|1 - \bar{a}_k r|^2} \right\} > 1 - \sum_{k=1}^{\infty} \frac{(1 - r^2)(1 - |a_k|^2)}{|1 - \bar{a}_k r|^2},$$

we may allow r to approach 1 under the summation sign, whence $\lim_{r \to 1} |B(r)| = 1$. Since the expression

$$\arg\left\{ 1 - \frac{1 - |a_k|^2}{1 - \bar{a}_k r} \right\}$$

is dominated by the general term of (2.7.5) for large k, it follows that the series
$$\arg B(r) = \sum_{k=1}^{\infty} \arg\left\{ 1 - \frac{1 - |a_k|^2}{1 - \bar{a}_k r} \right\}$$

converges absolutely and uniformly. This proves the first part of the theorem.

We assume, conversely, that every partial product of $B(z)$ tends to a limit of modulus 1 as z tends radially to 1. This implies that there can be only a finite number of zeros of $B(z)$ in any triangle lying inside $|z| < 1$ except for a vertex at $z = 1$. For otherwise a path terminating at $z = 1$ through an infinity of zeros in such a triangle would not be an asymptotic path for the corresponding partial product, contrary to Theorem 2.4. Thus, if we set $a_k = \alpha_k + i\beta_k$, we must have

$$\frac{|\beta_k|}{|1 - a_k|} > \sin \phi > 0,$$

and
$$\arg\left\{ 1 - \frac{1 - |a_k|^2}{1 - \bar{a}_k r} \right\} = \arcsin \frac{\beta_k r(1 - |a_k|^2)}{|a_k| \cdot |a_k - r| \cdot |1 - \bar{a}_k r|}$$

for each zero a_k in some neighbourhood N_ϵ: $|z - 1| < \epsilon$ of $z = 1$. We consider separately the two classes of zeros of $B(z)$: those of positive imaginary part and those of negative imaginary part. Since the partial sums of the arguments corresponding to each

class of zeros form a convergent series of positive (or negative) functions of r each of which is continuous in the closed interval $[0, 1]$, the convergence is uniform in this interval, by a known theorem (Hobson [1], vol. II, p. 116). We may therefore apply the inequality

$$\lim_{r \to 1} \arcsin \frac{\beta_k r(1 - |a_k|^2)}{|a_k| \cdot |a_k - r| \cdot |1 - \bar{a}_k r|}$$

$$= \arcsin \frac{\beta_k(1 - |a_k|^2)}{|a_k| \cdot |1 - a_k|^2} > \frac{d_k}{|1 - a_k|} \sin \phi$$

to the partial sums corresponding to points $a_k \in N_\varepsilon$ at $r = 1$. This, together with (2.6.3) for the points $a_k \notin N_\varepsilon$, implies the convergence of (2.7.5), which completes the proof of Theorem 2.13.

8. Further properties of Blaschke products. A uniqueness theorem

Each of the factors of a Blaschke product gives a one-to-one conformal mapping of $|z| \leqslant 1$ onto $|w| \leqslant 1$, and a finite Blaschke product gives a $(m, 1)$ conformal mapping of $|z| \leqslant 1$ onto $|w| \leqslant 1$. We prove next a theorem† which shows that, to a certain extent, i.e. except possibly for a set of capacity zero, this property is preserved by the infinite products; Theorem 2.14 will then be shown to be best possible in the sense that a Blaschke product can actually fail to assume a set of values in $|w| < 1$ of capacity zero. Since a Blaschke product has radial limit values of modulus 1 almost everywhere on $|z| = 1$, we state Theorem 2.14 for a more general class of functions; the significance of this wider class of functions will be felt in Chapter 5.

THEOREM 2.14. *Every bounded analytic function* $w = f(z)$ *in* $|z| < 1$ *having radial limits of modulus 1 almost everywhere on* $|z| = 1$ *and not reducing to a finite Blaschke product or a constant, assumes infinitely often every value of* $|w| < 1$ *except for a possible set of capacity zero.*

We assume, to the contrary, that there exists a closed set E of positive capacity lying in $|w| < \rho_0 < 1$, the values of which are not assumed infinitely often by $w = f(z)$. Thus there exists a

† Frostman [1].

distribution $\mu(\zeta)$ of unit mass on E such that the logarithmic potential

$$\int_{\zeta \in E} \log \frac{1}{|w-\zeta|} d\mu(\zeta)$$

is bounded in $|w| < 1$. It follows then that the potential

$$u(w) = \int_{\zeta \in E} \log \left| \frac{1-\bar{\zeta}w}{w-\zeta} \right| d\mu(\zeta) \qquad (2.8.1)$$

is also harmonic and bounded in $|w| < 1$ and, in addition, vanishes continuously on $|w| = 1$. Setting $w = f(r e^{i\theta})$ and computing the mean value of $u(f(z))$, we have

$$\frac{1}{2\pi} \int_0^{2\pi} u(f(r e^{i\theta})) \, d\theta$$
$$= \int_{\zeta \in E} \left\{ \frac{1}{2\pi} \int_0^{2\pi} \log \left| \frac{1-\bar{\zeta}f(r e^{i\theta})}{f(r e^{i\theta})-\zeta} \right| d\theta \right\} d\mu(\zeta).$$

From the definition of $u(w)$ and the fact that $u(f(r e^{i\theta})) \to 0$ as $r \to 1$ for almost all θ, $u(w)$ being bounded, we have

$$\lim_{r \to 1} \frac{1}{2\pi} \int_0^{2\pi} u(f(r e^{i\theta})) \, d\theta = \frac{1}{2\pi} \int_0^{2\pi} u(f(e^{i\theta})) \, d\theta = 0,$$

so that $\displaystyle \int_{\zeta \in E} \left\{ \lim_{r \to 1} \frac{1}{2\pi} \int_0^{2\pi} \log \left| \frac{1-\bar{\zeta}f(r e^{i\theta})}{f(r e^{i\theta})-\zeta} \right| d\theta \right\} d\mu(\zeta) = 0.$

Since the expression in parentheses of the integrand of this last integral is non-negative, it follows that, for all ζ in $|w| < 1$, except for a possible set of capacity zero,

$$\lim_{r \to 1} \int_0^{2\pi} \log \left| \frac{1-\bar{\zeta}f(r e^{i\theta})}{f(r e^{i\theta})-\zeta} \right| d\theta = 0.$$

By Theorem 2.12, the function $[f(z)-\zeta]/[1-\bar{\zeta}f(z)]$ must therefore be a Blaschke product $B(z)$ for all ζ in $|w| < 1$, except possibly for a set of capacity zero. Now if $B(z)$ has only a finite number of zeros, that is, if $f(z) = \zeta$ at only a finite number of points, then $f(z)$ is itself a finite Blaschke product, contrary to hypothesis, while if $B(z)$ has no zeros at all, then both $B(z)$ and $f(z)$ must be constant. Thus the only alternative is that $B(z)$, and *a fortiori* $f(z) - \zeta$, must have infinitely many zeros in $|z| < 1$, which proves Theorem 2.14.

In the course of the proof of Theorem 2.14, we have also proved the following useful result.†

THEOREM 2.15. *If $w = f(z)$ is analytic and bounded in $|z| < 1$ with radial limits of modulus 1 almost everywhere, then, for all ζ in $|w| < 1$ except possibly for a set of capacity zero, the function*

$$\frac{f(z) - \zeta}{1 - \bar{\zeta} f(z)} \tag{2.8.2}$$

is a Blaschke product.

To show that a Blaschke product can omit a non-countable set of values of capacity zero, we show first that Theorem 2.14 is best possible in the sense that, to an arbitrary closed set E_w of capacity zero in $|w| < 1$ there exists a function satisfying the hypotheses of Theorem 2.14 and omitting in $|z| < 1$ every value of E_w. Indeed, let E_w be an arbitrary closed set of capacity zero lying in $|w| < 1$, and let R_G denote the universal covering surface of the domain G obtained by deleting the closed set E_w from $|w| < 1$. If $z = z(w)$ denotes a function mapping R_G conformally onto $|z| < 1$, then the inverse function $w = f(z)$ of $z(w)$ is analytic and bounded, $|f(z)| < 1$, in $|z| < 1$, and assumes infinitely often every value of G but no value of E_w. Now if $f(z)$ does not satisfy the hypotheses of Theorem 2.14, then $f(z)$ must possess radial limit values of modulus less than 1 on a set of positive measure on $|z| = 1$. These radial limit values of modulus less than 1 must lie in E_w since all points of G correspond to interior points of R_G. To force our contradiction, we prove the following assertion. *If, on a set of positive measure on $|z| = 1$, the radial limit values $f(e^{i\theta})$ lie in a closed set E_w of capacity zero, then $f(z)$ is identically constant.* For, by Theorem 1.7, there exists a distribution $\mu(\zeta)$ of unit mass on E_w such that the potential

$$u(w) = \int_{\zeta \in E_w} \log \frac{1}{|w - \zeta|} \, d\mu(\zeta)$$

tends to $+\infty$ as w approaches any point of E_w. Hence the function

$$u_1(w) = \int_{E_w} \log \frac{2}{|w - \zeta|} \, d\mu(\zeta)$$

† Frostman [1].

is harmonic and non-negative in $|w| < 1$ and tends to $+\infty$ as w approaches any point of E_w. Thus the function $U(z) = u_1(f(z))$ is harmonic and non-negative in $|z| < 1$ and tends to $+\infty$ along a set of radii in $|z| < 1$ terminating in a set of positive measure on $|z| = 1$. If $V(z)$ denotes a harmonic conjugate of $U(z)$, then $F(z) = e^{-U-iV}$ is analytic and bounded in $|z| < 1$ with $F(e^{i\theta}) = 0$ on a set of positive measure. By Theorem 2.5, $F(z)$ must be identically zero; this contradiction then shows that $f(z)$ has radial limits of modulus 1 almost everywhere on $|z| = 1$, so that Theorem 2.14 is best possible. Theorem 2.14 shows that we may choose ζ in such a way that the function (2.8.2) is a Blaschke product $B(z)$; if $f(z)$ omits the set E_w of capacity zero, then $B(z)$ must omit the set which is the image of E_w under the transformation $(w-\zeta)/(1-\bar{\zeta}w)$.

A small modification of the argument used above gives us the following generalization of Theorem 2.5.

THEOREM 2.16. *If, on a set of positive measure on $|z| = 1$, the radial limit values of a bounded analytic function $w = f(z)$ lie in a set of capacity zero, then $f(z)$ is identically constant.*

For Egorov's theorem implies that, on a closed set of positive measure, the radial limits $f(e^{i\theta})$ lie in a closed set E_w of capacity zero. The potential

$$u_1(w) = \int_{E_w} \log \frac{2M}{|w-\zeta|}\, d\mu(\zeta),$$

where M is some bound on $|f(z)|$ in $|z| < 1$, replaces the function $u_1(w)$ used above, and the remaining details are exactly as before.†

9. Functions of bounded characteristic and functions omitting three values

Hitherto in this chapter we have considered only bounded functions, which are necessarily regular. We now extend the investigation to functions which are meromorphic in $|z| < 1$. To do this, we recall some general theorems on meromorphic func-

† For more general theorems related to Theorem 2.16, see Nevanlinna [1], pp. 202 ff.

tions. The standard notation of the Nevanlinna theory† will be used. We shall write

$$T(r,f) = m(r,f) + N(r,f),$$

where

$$m(r,f) = \frac{1}{2\pi} \int_0^{2\pi} \log^+ |f(re^{i\theta})| \, d\theta,$$

and

$$N(r,f) = \int_0^r \frac{n(t,\infty) - n(0,\infty)}{t} \, dt + n(0,\infty) \log r,$$

where $n(t,\infty)$ denotes the number of poles of $f(z)$ in $|z| \leqslant t$, each being counted according to its multiplicity, and where

$$\log^+ |f(re^{i\theta})| = \log |f(re^{i\theta})|$$

for all θ for which $\log |f| > 0$ and is zero elsewhere. More generally $n(t,a)$ is the number of zeros of $f(z) - a$ in $|z| \leqslant t$ counted according to multiplicity, and

$$N(r,a) = \int_0^r \frac{n(t,a) - n(0,a)}{t} \, dt + n(0,a) \log r.$$

If the values of $f(z)$ are transformed to the unit w-sphere, the function $T(r,f)$ must, of course, be defined in terms of the spherical metric.

The function $T(r,f)$ is non-decreasing and, for a function meromorphic in $|z| < 1$, may be either bounded, in which case $f(z)$ is said to be *of bounded characteristic*, or unbounded. Further,

$$T\left(r, \frac{\alpha f + \beta}{\gamma f + \delta}\right) = T(r,f) + O(1), \tag{2.9.1}$$

where $(\alpha f + \beta)/(\gamma f + \delta)$ is a linear transform of f, so that if $f(z)$ is of bounded characteristic, so is $(\alpha f + \beta)/(\gamma f + \delta)$.

A function $f(z)$ meromorphic in $|z| < 1$ can be expressed as a quotient of two functions $\phi_1(z)$, $\phi_2(z)$, both regular and bounded in $|z| < 1$, if and only if $T(r,f)$ is bounded.

In consequence of this theorem of Nevanlinna, *functions of bounded characteristic*, or, as they are sometimes called, *functions of bounded type*, share many of the properties of bounded functions, including the Fatou property mentioned at the end of §4. There are, however, important exceptions, as we shall see later.

† Nevanlinna [1].

In §3 we showed that if α is an *asymptotic value* of a bounded analytic function at $e^{i\theta_0}$, that is, if $f(z) \to \alpha$ as $z \to e^{i\theta_0}$ along a path in $|z| < 1$ terminating at $e^{i\theta_0}$, then α is a radial limit of $f(z)$ at $e^{i\theta_0}$. Moreover, Theorem 2.3 states that $f(z)$ tends to α uniformly inside any angle with sides in $|z| < 1$ and vertex at $e^{i\theta_0}$. We point out at once that this theorem, often called Lindelöf's theorem, does not admit an immediate extension to an arbitrary class of unbounded analytic functions, as the following example shows. We have already observed that the function $e^{(z+1)/(z-1)}$ is analytic and bounded in $|z| < 1$ and tends to 0 as $z \to 1$ along the radius drawn to $z = 1$. In addition, the modulus of $e^{(z+1)/(z-1)}$ is constant, namely e^{-1}, on the circle $|z - \tfrac{1}{2}| = \tfrac{1}{2}$, so that, as $z \to 1$ along $|z - \tfrac{1}{2}| = \tfrac{1}{2}$, this function has a constant, non-zero modulus. Therefore the function

$$\phi(z) = (z-1)\, e^{(1+z)/(1-z)} \tag{2.9.2}$$

has the property that it tends to zero as $z \to 1$ along the path $|z - \tfrac{1}{2}| = \tfrac{1}{2}$, yet the radial limit is not zero but ∞.

The function defined in (2.9.2) belongs to a special class of functions to which the theorems of Fatou and Riesz may be extended, namely, functions of bounded characteristic in $|z| < 1$, which, by the theorem of Nevanlinna quoted above, can be expressed as the quotient of two bounded regular functions. Theorem 2.10 of §6 may be carried over immediately to this class of meromorphic functions, and we have

THEOREM 2.17. *A necessary and sufficient condition that a meromorphic function be of bounded characteristic in $|z| < 1$ is that it have an integral representation of the form*

$$f(z) = e^{i\alpha} \frac{B_1(z)}{B_2(z)} \exp\left\{ \frac{1}{2\pi} \int_0^{2\pi} \frac{e^{it}+z}{e^{it}-z}\, d\mu(t) \right\}, \tag{2.9.3}$$

where $\mu(\theta)$ is of bounded variation in $0 \leqslant \theta \leqslant 2\pi$, α is a constant, and where $B_1(z)$ and $B_2(z)$ are Blaschke products extended over the zeros $\{a_k\}$ and poles $\{b_j\}$ respectively of $f(z)$ with

$$\sum_{k=1}^{\infty}(1 - |a_k|) < \infty; \quad \sum_{j=1}^{\infty}(1 - |b_j|) < \infty. \tag{2.9.4}$$

If $f(z)$ is of bounded characteristic in $|z| < 1$, and if

$$f(z) = f_1(z)/f_2(z),$$

where $f_1(z)$ and $f_2(z)$ are analytic and bounded in $|z| < 1$, it follows from Theorem 2.1 that $f_1(z)$ and $f_2(z)$ possess radial limits almost everywhere on $|z| = 1$. Since $\liminf_{r \to 1} |f_2(r\,e^{i\theta})| = 0$ on at most a set of measure zero, according to Theorem 2.5, it follows that the radial limits of $f(z)$ exist for almost all $e^{i\theta}$ on $|z| = 1$, so that we have

THEOREM 2.18. *A function $f(z)$, which is meromorphic and of bounded characteristic in $|z| < 1$, possesses radial limits at almost all points of $|z| = 1$.*

Suppose next that the radial limit values of a function $f(z)$, of bounded characteristic in $|z| < 1$, are zero on a set of positive measure on $|z| = 1$. Since $T(r,f)$ is bounded, $m(r,a)$, defined below, is bounded for all a. If now $f(r\,e^{i\theta}) \to a \neq \infty$ for $\theta \in E$, $mE > 0$, then by Egorov's theorem, $f(r\,e^{i\theta}) \to a$ uniformly in a set $E_1 \subset E$, where $mE_1 > 0$. Therefore

$$m(r,a) = m\left(r, \frac{1}{f-a}\right) = \frac{1}{2\pi}\int_0^{2\pi} \log^+ \frac{1}{|f(r\,e^{i\theta}) - a|}\, d\theta$$

$$> \frac{1}{2\pi}\int_{E_1} \log^+ \frac{1}{|f(r\,e^{i\theta}) - a|}\, d\theta \to \infty \quad \text{as} \quad r \to 1,$$

so that, by (2.9.1), $T(r,f) > m(r,a) - O(1)$ is unbounded, contrary to hypothesis. We have thus proved

THEOREM 2.19. *A function $f(z)$ which is of bounded characteristic in $|z| < 1$ and whose radial limits are equal to a constant on a set of positive measure on $|z| = 1$ is identically constant in $|z| < 1$.*

Since a bounded function is of bounded characteristic, Theorem 2.5 is contained in Theorem 2.19.

Thus the theorems of Fatou and Riesz may be extended to functions of bounded characteristic,† which may be shown to possess angular limits almost everywhere on $|z| = 1$. The function (2.9.2) shows, however, that Theorem 2.3 cannot be extended to this class of functions. On the other hand, there is an important

† Nevanlinna [1], pp. 185 ff.

class of functions, namely those which omit three values to which Theorem 2.3 may be extended.

To prove this we first recall the following lemma due to Koebe.

LEMMA. *Let $f(z)$ be analytic and bounded in $|z| < 1$ and let there be a sequence of arcs γ_n each contained in an annulus $1 - \epsilon_n < |z| < 1$, $\lim_{n\to\infty} \epsilon_n = 0$, having as limiting set an arc γ of $|z| = 1$ and such that on γ_n, $|f(z) - a| < \eta_n$, $\lim_{n\to\infty} \eta_n = 0$. Then $f(z) \equiv a$.*

To exploit this lemma we need two further definitions:

(i) $\Phi(f)$ is the set of values a such that there exists a sequence $\{\eta_n\}$ of arcs satisfying the conditions of the lemma and on which $f(z) \to a$ uniformly;

(ii) $\Phi(f, e^{i\theta})$ is the set of values a such that $e^{i\theta}$ belongs to some closed arc γ of $|z| = 1$ satisfying the conditions of the lemma.

We now prove

THEOREM 2.20. *Let $f(z)$ be meromorphic and non-constant in $|z| < 1$, and let $f(z)$ omit three distinct values a, b, and c in $|z| < 1$. Then, (i) if α is an asymptotic value along some path \mathscr{L} terminating at a point $e^{i\theta_0}$, then $f(z)$ tends to α uniformly as $z \to e^{i\theta_0}$ inside any angular domain lying in $|z| < 1$ and having $e^{i\theta_0}$ as vertex; and (ii) $\Phi(f)$ is empty.*

We assume that $f(z)$ has been transformed so as to make $a = 0$, $b = 1$, and $c = \infty$, so that the omitted values are 0, 1, and ∞. Now if $\nu(w)$ denotes the inverse of the modular function for the half-plane, the function

$$F(z) = \frac{\nu(f(z)) - \nu(i)}{\nu(f(z)) - \overline{\nu(i)}},$$

for a given branch of the multiform function $\nu(w)$, is analytic and bounded in $|z| < 1$, $|F(z)| < 1$, and satisfies the hypotheses of Theorem 2.3, for, as $z \to e^{i\theta_0}$ along \mathscr{L}, $F(z)$ tends to a limit, say β, so that $F(z)$ tends uniformly to β inside any angular domain in $|z| < 1$ with vertex at $e^{i\theta_0}$. It follows that $f(z) \to \alpha$ as $z \to e^{i\theta_0}$ in this same angular domain.†

To prove, (ii), suppose there is a Koebe sequence $\{\gamma_n\}$ of curves in $|z| < 1$ and a sequence of numbers w_n such that $|f(z) - w_n| < \epsilon_n$ on γ_n, $\lim_{n\to\infty} \epsilon_n = 0$ and $\lim_{n\to\infty} w_n = d \in \Phi(f)$. For a

† Collingwood and Cartwright [1].

given branch of $\nu(w)$ the sequence $\dfrac{\nu(w_n) - \nu(i)}{\nu(w_n) - \nu(\bar{i})}$ tends to a limit β' so that on γ_n, $|F(z) - \beta'| < \eta_n$ where $\lim\limits_{n \to \infty} \eta_n = 0$. Hence $F(z) \equiv \beta'$ and $f(z) =$ constant, contrary to hypothesis.

Theorem 2.20 implies the following theorem of Lindelöf [1]:

THEOREM 2.21. *If $f(z)$ is meromorphic in $|z| < 1$, and if $f(z)$ admits two distinct asymptotic values at some point $e^{i\theta_0}$, then $f(z)$ assumes infinitely often in any neighbourhood of $e^{i\theta_0}$ all values of the extended complex plane with at most two possible exceptions.*

A point $e^{i\theta_0}$ on $|z| = 1$ such that $f(z)$ assumes infinitely often in any vicinity of $e^{i\theta_0}$ all complex values with at most two exceptions will be called a *Picard point* of $f(z)$; any such exceptional value will be called a *Picard value* of $f(z)$.

10. Proof that Fatou's theorem is best possible

We conclude this chapter by showing that the form of Fatou's theorem given in Theorem 2.4 is best possible in the sense that it cannot be extended further to give limiting values along paths lying outside every angular domain with vertex on $|z| = 1$, and by showing that Theorem 2.7 allows us to give a simple construction of a function meromorphic in $|z| < 1$ having *no* radial limits.

THEOREM 2.22.† *Let γ_0 be a simple closed Jordan curve, internally tangent to $|z| = 1$ at $z = 1$, and having no other points in common with $|z| = 1$; let γ_θ denote the rotation of γ_0 through an angle θ about the origin. Then there exists a function $f(z)$, analytic and bounded in $|z| < 1$, which does not approach a limit as z approaches any point $e^{i\theta}$, from the right or the left, along γ_θ.*

Since the curve γ_0 is tangent to $|z| = 1$ at $z = 1$, there exists a simple closed curve γ_1 which lies inside of γ_0 except for the point $z = 1$, and which has an equation of the form $\theta = \pm \lambda(r)$, where $\lambda(r)$ is a monotonely decreasing function in the neighbourhood of $r = 1$, and where $(1-r)/\lambda(r) \to 0$. We choose a sequence $\{r_n\}$ such that

$$\sum_{n=1}^{\infty} (1 - r_n)/\lambda(r_n) < \infty,$$

† Lohwater and Piranian [1]. A weaker version of this theorem, in which the function possesses the asserted property *almost everywhere* on $|z| = 1$ was given earlier by Littlewood [1]; a simpler proof of Littlewood's example, again with the property holding *almost everywhere* on $|z| = 1$, was given by Zygmund [1].

and define the functions

$$f_n(z) = \frac{z^{k_n} - r_n^{k_n}}{(r_n z)^{k_n} - 1},$$

where k_n is that integer such that

$$\frac{3\pi}{\lambda(r_n)} < k_n \leqslant \frac{3\pi}{\lambda(r_n)} + 1.$$

The infinite product $\prod\limits_{n=1}^{\infty} f_n(z)$ converges because

$$\sum_{n=1}^{\infty} k_n(1 - r_n) < \infty.$$

Also, for each fixed r,

$$\lim_{n \to \infty} f_n(r\,e^{i\theta}) = \lim_{n \to \infty} r_n^{k_n} = 1$$

uniformly with respect to θ, while, for each fixed n,

$$\lim_{r \to 1} |f_n(r\,e^{i\theta})| = 1$$

uniformly with respect to θ. Thus, if the sequence $\{n_p\}$ increases rapidly enough, the function $f(z) = \prod\limits_{p=1}^{\infty} f_{n_p}(z)$ has the property that $|f(z)| > \frac{1}{2}$ everywhere on each of an expanding sequence of circles $|z| = \rho_p$ with $\lim\limits_{p \to \infty} \rho_p = 1$. Since the function $f_n(z)$ has k_n uniformly spaced zeros on the circle $|z| = r_n$, it follows that for each point $e^{i\alpha}$ there exists a path, lying between the radius $\theta = \alpha$ and the left half of the curve γ_1, along which $f(z)$ does not approach a limit. Because of the presence of zeros of $f(z)$ in this region, it follows from Theorem 2.3 that if $f(z)$ approached a limit along the left-hand parts of γ_0 or γ_1, this limit would have to be 0, which would then contradict the fact that $|f(z)| > \frac{1}{2}$ on the circles $|z| = \rho_p$. Since the same argument applies to the right-hand portions of these curves, the proof is complete.

Theorem 2.7 shows us how we may construct functions having no radial limit[†] (in the extended sense) at any point of $|z| = 1$. As an example, we consider the modular function $\mu(z)$, which is

[†] Lohwater and Piranian [1]. It has been shown (Lehto and Virtanen [1]) that the example given here possesses no asymptotic values. This distinguishes it from the functions having more than one asymptotic value on spiral paths which also have no radial limit values (see Chapter 6).

analytic in $|z| < 1$ and omits the values 0, 1 and ∞. If we transform $\mu(z)$ so that the three omitted values are finite, say a, b, and c, then the transformed function $g(z)$, like $\mu(z)$, possesses radial limits on a countable set E; these radial limits are, of course, the values a, b and c. Since E is countable, it is a set of type F_σ, so that, by Theorem 2.7, there exists a function $h(z)$, analytic and bounded in $|z| < 1$, such that E is precisely the set on which $h(z)$ fails to have a radial limit. The function $f(z) = g(z) + h(z)$ then cannot have a radial limit at any point of $|z| = 1$.

Since the function $g(z)$ is obtained from the modular function $\mu(z)$ by a linear fractional transformation, it follows that

$$T(r, g) = O\left(\log\frac{1}{1-r}\right),$$

and since $T(r, h) = O(1)$, it follows that

$$T(r, f) = O\left(\log\frac{1}{1-r}\right).$$

The construction of the function $f(z)$ thus answers negatively the question of whether a function whose characteristic function is of order no greater than $O(\log 1/(1-r))$ must possess radial limit values on a set which is dense on $|z| = 1$.

We remark that MacLane [1] has shown recently that there exist in $|z| < 1$ meromorphic functions without asymptotic values such that the characteristic function $T(r)$ increases arbitrarily slowly. For a more comprehensive survey, see MacLane [4].

It will be appreciated that in this chapter we have by no means exhausted the topic of Fatou's theorem, which is a focal point of the theory of cluster sets in the disc. Generalizations of the theorem will be discussed in Chapter 8; but one particular generalization due to Meier [1]† may be mentioned here although we have not space to give the proof and must refer the reader to Meier's paper or Noshiro's book. The theorem in question is the following:

THEOREM 2.23. *If $f(z)$ is analytic in $|z| < 1$ and if E is a measurable set on $|z| = 1$ of positive measure such that at every point $e^{i\theta}$ of E the function $f(z)$ is bounded on two distinct chords terminating at $e^{i\theta}$, then almost all points of E are Fatou points of $f(z)$.*

† See also Noshiro [4].

TOPICS IN THE THEORY OF CONFORMAL MAPPING

1. Introduction

Many of the results of Chapter 2, proved for the unit circle or unit disc, may be given a more general setting by means of conformal mapping. Since our theory is, in the main, a theory of boundary behaviour, the results which we require from the theory of conformal mapping for this purpose are those relating to the correspondence of frontiers under such a mapping and not the 'interior' properties. Such theorems are of two types, metrical and topological. Those relating to the mapping of a Jordan domain onto the unit disc and the metrical problems involved in the boundary correspondence are of special importance. In this chapter we shall therefore restrict ourselves to the Jordan domain. The theory of the correspondence between the boundary of the general simply connected domain and the unit circle under a conformal mapping involves the topological notion of a prime end and will be developed in Chapter 9.

2. Boundary correspondence under conformal mapping of a Jordan domain

By definition, a Jordan curve is a homeomorphism, or continuous one-to-one mapping, of a circle, and a Jordan domain is a domain whose boundary is a Jordan curve. Our first problem is to show that a conformal mapping of the open disc $|z| < 1$ onto a Jordan domain D may be extended to a homeomorphism between $|z| \leqslant 1$ and the closure \overline{D} of D. The extension is effected by setting $f(p) = C(f, p)$, where $p \in K \colon |z| = 1$, and showing that when D is a Jordan domain the cluster set $C(f, p)$ is degenerate, i.e. consists of a single point, and is a continuous function of p. (Trivially, we observe that for $|z| < 1$ we have $f(z) = C(f, z)$, so that the extension is a perfectly natural one. We shall use the same extension whenever we consider the mapping of $|z| < 1$ onto

a general domain, but in this case the cluster set $C(f, p)$ will not necessarily be degenerate for $p \in K$, so that a different treatment is necessary.)

Our development is based on the following theorem of Tsuji which requires a few preliminaries in order to state it. We take p to be the point $z = 1$ and, for $m = 1, 2, \ldots$, we denote by G_m the intersection of U: $|z| < 1$ with the annulus $2^{-m} \geqslant |z-1| \geqslant 2^{-m-1}$ and by $L(r)$ the length of the image under the mapping $f(z)$ of the intersection of U with the circle $|z-1| = r$. Further, we write

$$L_m = \inf_{2^{-m} \geqslant r \geqslant 2^{-m-1}} L(r).$$

We now prove

LEMMA 1. *If $f(z)$ maps U conformally onto a bounded domain, then, in the above notation,* $\qquad L_m = o(1).$ $\qquad\qquad$ (3.2.1)

Note that, in this lemma, the domain is not restricted to be a Jordan domain.

Let D be the domain. Since D is of finite area and contains the images of the domains G_m, it follows that

$$\iint_{G_m} |f'|^2 \, d\sigma$$

tends to zero as $m \to \infty$. Hence, by Schwarz's inequality,

$$\left(\iint_{G_m} |f'| \, d\sigma \right)^2 \leqslant \left(\iint_{G_m} |f'|^2 \, d\sigma \right) \left(\iint_{G_m} d\sigma \right) = o(4^{-m}),$$

and, since $\qquad\qquad 2^{-m-1} L_m \leqslant \iint_{G_m} |f'| \, d\sigma,$

the estimate (3.2.1) follows.

We denote by γ_m a circular arc of centre $z = 1$ contained in G_m, having end-points a_m, b_m on K, and such that its image $f(\gamma_m)$ is of length L_m. Thus $f(\gamma_m)$, being of finite length, is a Jordan arc in D whose end-points α_m, β_m lie on FR D.

At the points a_m, b_m, the function $f(z)$ has the Fatou values (angular limits) α_m, β_m. By Theorem 2.5, the set E_m of points of K at which $f(z)$ has any of the Fatou values α_ν, β_ν, $\nu < m$, is of measure zero. Therefore, since

$$\iint_{G_m'} |f'(z)| \, d\sigma \leqslant \iint_{G_m} |f'(z)| \, d\sigma = o(1),$$

where G'_m is obtained by deleting from G_m the circular arcs of centre $z = 1$ whose end-points include a point of E_m, we have

$$L'_m = o(1),$$

where L'_m denotes the infimum of $L(r)$ in G'_m. Continuing the process for all finite m, we see that there exists a sequence of circular arcs $\gamma_m \subset G_m$, $m = 1, 2, \ldots$, such that the corresponding curves $f(\gamma_m)$ under the mapping by f are cross-cuts of D whose lengths tend to zero as $n \to \infty$ and no one of which has an end-point in common with another. Accordingly, $\overline{f(\gamma_m)} \cap \overline{f(\gamma_{m+k})} = \varnothing$ for all $k > 0$.

We have thus proved

THEOREM 3.1. *If $f\{z\}$ maps $|z| < 1$ conformally onto a bounded domain D, there exists a sequence of circular arcs $\gamma_\nu : \{z : |z - 1| = r_\nu\}$, $|z| < 1$, where $r_1 > r_2 > \ldots > r_\nu > \ldots$, $\lim\limits_{\nu \to \infty} r_\nu = 0$, such that their images $f(\gamma_\nu)$ form a sequence of disjoint Jordan arcs in D having their end-points on $\mathrm{FR}\, D$, their closures being pairwise disjoint, and such that each of the arcs separates all its successors in the sequence from its predecessor, while the length of $f(\gamma_\nu)$ tends to zero as $\nu \to \infty$.*

We remark again that D is not necessarily a Jordan domain. The circular arcs γ_ν are *cross-cuts* of $|z| < 1$ and their images $f(\gamma_\nu)$ are cross-cuts of D. Both $\{\gamma_\nu\}$ and $\{f(\gamma_\nu)\}$ form what we shall define in Chapter 9 as chains of cross-cuts in $|z| < 1$ and D respectively.

Suppose now that $\mathrm{FR}\, D$ is a Jordan curve C, let δ_ν be that subdomain of $|z| < 1$ cut off by γ_ν and containing $\gamma_{\nu+1}$, and denote by $d_\nu = f(\delta_\nu)$ the image of δ_ν under the mapping. The boundary of d_ν consists of the cross-cut $f(\gamma_\nu)$ of D and an arc A_ν of C having end-points α_ν and β_ν. Since the diameter of $f(\gamma_\nu)$ tends to zero and since C is a Jordan curve, the diameter of A_ν, and consequently also the diameter of d_ν, tends to zero as $\nu \to \infty$. Hence the cluster set $C(f, 1) = \bigcap\limits_\nu \overline{d}_\nu \subset C$ is of zero diameter and so consists of a single point which we may denote by $f(1)$, since, for any sequence $\{z_n\}$, $|z_n| < 1$, converging to $z = 1$, $\lim\limits_{n \to \infty} f(z_n) = f(1)$.

For two points $e^{i\theta}$, $e^{i\varphi}$, $\theta \neq \varphi$, we can construct sequences of cross-cuts $\gamma_\nu(\theta)$, $\gamma_\nu(\varphi)$ of $|z| < 1$ cutting off subdomains $\delta_\nu(\theta)$,

$\delta_\nu(\varphi)$ as described above, the corresponding cross-cuts and sub-domains of D under the mapping f being $f(\gamma_\nu(\theta))$, $f(\gamma_\nu(\varphi))$ and $d_\nu(\theta)$, $d_\nu(\varphi)$, respectively. For sufficiently large ν the domains $\delta_\nu(\theta)$, $\delta_\nu(\varphi)$ are at a positive distance apart, from which it follows that the domains $d_\nu(\theta)$, $d_\nu(\varphi)$ do not meet. We can also assert that the closures $\overline{d_\nu(\theta)}$, $\overline{d_\nu(\varphi)}$ do not meet. For the boundary of $d_\nu(\theta)$ consists of the closure $\overline{f(\gamma_\nu(\theta))}$ of $f(\gamma_\nu(\theta))$ and an open arc $A_\nu(\theta)$ of C, while the boundary of $d_\nu(\varphi)$ consists of $\overline{f(\gamma_\nu(\varphi))}$ and an open arc $A_\nu(\varphi)$ of C. Since C is a Jordan curve and $d_\nu(\theta) \cap d_\nu(\varphi) = \varnothing$, $A_\nu(\theta)$ does not meet $A_\nu(\varphi)$. Further, $f(\gamma_\nu(\theta))$ does not meet $f(\gamma_\nu(\varphi))$. It remains to consider the end-points of these two cross-cuts. If $f(\gamma_\nu(\theta))$ and $f(\gamma_\nu(\varphi))$ have a common end-point then, by our construction, for all $k > 0$, $f(\gamma_{\nu+k}(\theta))$ and $f(\gamma_{\nu+k}(\varphi))$ have their end-points in $A_\nu(\theta)$ and $A_\nu(\varphi)$ respectively so that

$$\overline{f(\gamma_{\nu+k}(\theta))} \cap \overline{f(\gamma_{\nu+k}(\varphi))} = \varnothing$$

and hence $$\overline{d_\nu(\theta)} \cap \overline{d_\nu(\varphi)} = \varnothing \,,$$

as we had to prove. It follows that $f(e^{i\theta}) \neq f(e^{i\varphi})$. The correspondence between points of $|z| = 1$ and C by the function

$$f(e^{i\theta}) = C(f, e^{i\theta})$$

is therefore one-to-one and continuous. We have thus proved

THEOREM 3.2. *If $w = f(z)$ maps $|z| < 1$ conformally onto a domain D bounded by a Jordan curve C, then $f(z)$ may be extended to $|z| \leqslant 1$ by defining $f(e^{i\theta})$ to be the (degenerate) cluster set $C(f, e^{i\theta})$ so as to give a mapping of $|z| \leqslant 1$ onto \overline{D}, which is a homeomorphism.*

3. The Riesz theorem on conformal mapping of a Jordan domain with a rectifiable boundary

The one-to-one continuous correspondence between C and $|z| = 1$ asserted by Theorem 3.2 assumes a stronger form, as might be expected, when additional restrictions are placed upon the Jordan curve C. The most important restriction which we shall consider here is that of rectifiability. With this condition, it was proved by F. and M. Riesz [1] that the boundary homeomorphism, which we denote by $w = f(e^{i\theta})$, is *absolutely continuous*

on $|z| = 1$, from which it follows that a set of measure zero on $|z| = 1$ corresponds to a set of measure zero on C.

We prove the Riesz theorem, namely,

THEOREM 3.3. *Let $w = f(z)$ be an analytic function mapping $|z| < 1$ conformally onto a domain D bounded by a rectifiable Jordan curve C. Then, under the homeomorphism $w = f(e^{i\theta})$ of the frontiers induced by the mapping $w = f(z)$, a set of measure zero on $|z| = 1$ is mapped onto a set of measure zero on C, and conversely. Further, if L be the length of C, then*

$$L = \int_0^{2\pi} |f'(e^{i\theta})|\, d\theta. \qquad (3.3.1)$$

We note expressly that if the rectifiability of C is dispensed with, then an example can be given† in which a set E on C of positive linear measure corresponds to a set E on $|z| = 1$ of measure zero; the set E lies on a straight line and its measure is defined with respect to that line, not with respect to C.

Since C is rectifiable the continuous function $F(\theta) = f(e^{i\theta})$ is of bounded variation on the interval $(0, 2\pi)$ and

$$F'(\theta) = \frac{d}{d\theta} F(\theta) = i\, e^{i\theta} f'(e^{i\theta})$$

exists and is finite almost everywhere.

The mapping function $f(z)$ maps $|z| < 1$ onto D and it maps $|z| = r$, $0 < r < 1$, onto a curve C_r whose length we denote by L_r. Then

$$L_r = \int_0^{2\pi} |f'(r\, e^{i\theta})|\, r\, d\theta.$$

Let $\omega = e^{2\pi i/n}$ and put

$$\Lambda_n(z) = |f(z) - f(\omega z)|$$
$$+ |f(\omega z) - f(\omega^2 z)| + \ldots + |f(\omega^{n-1} z) - f(z)|.$$

The function $\Lambda_n(z)$, being a finite sum of moduli of functions each of which is regular in $|z| < 1$ and continuous in $|z| \leqslant 1$, is subharmonic‡ in $|z| < 1$ and continuous in $|z| \leqslant 1$. $\Lambda_n(z)$ therefore

† Lohwater and Seidel [1]. An earlier example was given by Lavrentiev [1]. See also Lohwater and Piranian [3] and, for recent developments, Matsumoto [5].

‡ For the theory of subharmonic functions see Littlewood [2] or Tsuji [4]. The result can also be deduced directly from the maximum modulus theorem for analytic functions, cf. Zygmund [3], p. 160.

attains its maximum at a point z_0 of $|z| = 1$ so that, putting $z = r$,

$$\Lambda_n(r) \leqslant \Lambda_n(z_0).$$

Since $\Lambda_n(z_0)$ is the length of a polygon whose vertices lie on C we have $\Lambda_n(z_0) \leqslant L$ and $\Lambda_n(r) \leqslant L$ for all $0 < r \leqslant 1$ and for all $n < \infty$. But

$$\lim_{n \to \infty} \Lambda_n(r) = L_r,$$

so that $L_r \leqslant L$, $0 < r < 1$, and so $f'(z) \in H^1$, where H^1 is the Hardy class of functions $\varphi(z)$ analytic in $|z| < 1$ such that

$$\int_0^{2\pi} |\varphi(r\, e^{i\theta})|\, d\theta = O(1).$$

This validates the Poisson integral representation†

$$izf'(z) = \frac{1}{2\pi} \int_0^{2\pi} i\, e^{it} f'(e^{it})\, P(r, t - \theta)\, dt$$

$$= \frac{1}{2\pi} \int_0^{2\pi} \frac{dg(t)}{dt} P(r, t - \theta)\, dt,$$

where $P(r, t - \theta)$ is the Poisson kernel

$$\frac{1 - r^2}{1 - 2r \cos(t - \theta) + r^2}, \quad z = r\, e^{i\theta},$$

and

$$g(t) = \int_0^t i\, e^{i\tau} f'(e^{i\tau})\, d\tau.$$

By partial integration we now have, since $g(2\pi) = g(0)$,

$$izf'(z) = \frac{1}{2\pi} \int_0^{2\pi} g(t) \frac{\partial P(r, t - \theta)}{\partial \theta}\, dt,$$

or

$$\frac{\partial f(r\, e^{i\theta})}{\partial \theta} = \frac{\partial}{\partial \theta} \frac{1}{2\pi} \int_0^{2\pi} g(t)\, P(r, t - \theta)\, dt,$$

so that

$$f(r\, e^{i\theta}) = \frac{1}{2\pi} \int_0^{2\pi} g(t)\, P(r, t - \theta)\, dt + c(r),$$

where $c(r)$ is a function depending only on r. Since the integral on the right is a (complex) harmonic function, $c(r)$ is harmonic and thus satisfies Laplace's equation

$$\Delta c(r) = \frac{d^2}{dr^2} c(r) + \frac{1}{r} \frac{d}{dr} c(r) = 0.$$

† A very full treatment is given in Tsuji [4] and Golusin [1].

Hence $c(r) = a \log r + b$, a and b being constants. But $c(r)$ is continuous at $r = 0$ so that $a = 0$ and $c(r) = b$ and we have

$$f(r\,e^{i\theta}) = \frac{1}{2\pi} \int_0^{2\pi} (g(t)+b)\,P(r,t-\theta)\,dt. \qquad (3.3.2)$$

Now $g(t)$, being an integral, is absolutely continuous on $(0, 2\pi)$. Also, by a well-known theorem, the Poisson integral on the right is a (complex) harmonic function which, since $g(t)$ is continuous, tends to $g(t) + b$ as $z \to e^{it}$ for all values of t. Therefore $f(e^{it}) = g(t) + b$ almost everywhere and since $f(e^{it})$ is continuous and $g(t) + b$ is absolutely continuous it follows that $f(e^{it}) = F(t)$ is absolutely continuous. Hence, by a theorem of Lebesgue, the length l of the arc of C corresponding to an arc α of $|z| = 1$ is given by

$$l = \int_\alpha |F'(\theta)|\,d\theta = \int_\alpha |f'(e^{i\theta})|\,d\theta. \qquad (3.3.3)$$

We are now in a position to prove our theorem.

We begin with an arbitrary set E_z of measure zero on $|z| = 1$; we shall show that the image E_w of E_z on C is also of measure zero. We find it convenient to show that we can restrict ourselves without loss of generality to the case that E_z, and therefore E_w, is closed. Since E_z is of measure zero, it can be covered by a system of closed intervals of arbitrarily small total length. From this system we can choose an infinite sequence of coverings of E_z, for which the total length tends to zero. Let E_z' denote the set of all points common to the whole system of coverings; clearly, E_z' contains E_z and is of measure zero. Now the image E_w' of E_z' is measurable because it is the intersection of the intervals which are the images of the intervals which form the coverings of E_z. Since E_w' contains E_w, it will suffice to prove that $m(E_w') = 0$. Now if E_w' were of positive measure, it would contain a closed subset of positive measure. Thus it suffices to assume that E_z and its image E_w on C are closed. If we cover E_z with a finite set G_z of open intervals of total length less than ϵ, the image G_w of G_z on C contains E_w, and we have, in virtue of (3.3.3),

$$m(E_w) \leqslant m(G_w) = \int_{G_z} |f'(e^{i\theta})|\,d\theta. \qquad (3.3.4)$$

Since the integral in (3.3.4) can be made arbitrarily small by making $m(G_z)$ small, we have shown that a set of measure zero on $|z| = 1$ is mapped into a set of measure zero on C.

Conversely, we wish to show that a set E_w of measure zero on C corresponds to a set of measure zero on $|z| = 1$. As before, we may assume that E_w is closed.

Now a function of class H^1 is of bounded type, for $\log^+ |\varphi| \leqslant |\varphi|$. Hence $f'(z)$ being of class H^1 is of bounded type and it follows from Theorem 2.19 that $f'(e^{i\theta})$ cannot vanish on $|z| = 1$ on a set of positive measure. Therefore the measure of the set of points on $|z| = 1$ where $|f'(e^{i\theta})| < \epsilon$ must tend to zero with ϵ, so that, if $\eta(\epsilon)$ denotes the measure of the set on $|z| = 1$ where $|f'(e^{i\theta})| < \epsilon$, then $\lim\limits_{\epsilon \to 0} \eta(\epsilon) = 0$. Now, if E is any measurable set on $|z| = 1$ for which $\int_E |f'(e^{i\theta})| \, d\theta < \epsilon$, we must have

$$\int_E |f'(e^{i\theta})| \, d\theta \geqslant \int_{E_1} |f'(e^{i\theta})| \, d\theta \geqslant \sqrt{(\epsilon)} \, (m(E) - \eta(\sqrt{\epsilon})),$$

where E_1 denotes that subset of E for which $|f'(e^{i\theta})| > \sqrt{\epsilon}$. Thus we have $m(E) < \sqrt{(\epsilon)} + \eta(\sqrt{\epsilon})$, from which it follows that

$$\lim_{\epsilon \to 0} m(E) = 0.$$

Now if the closed set E_w on C is of measure zero, it can be covered by a finite set G_w of arcs of C of total length less than ϵ. The arcs G_w on C are the images of a set G_z of arcs on $|z| = 1$, and we have

$$m(G_w) = \int_{G_z} |f'(e^{i\theta})| \, d\theta < \epsilon,$$

from which it follows that $m(G_z) < \sqrt{(\epsilon)} + \eta(\sqrt{\epsilon})$. Hence $m(G_z) = 0$, and Theorem 3.3 is proved.

In order to prove that the image E_w on C of a set E_z of positive measure on $|z| = 1$ is of positive measure, it suffices merely to prove that E_w is measurable. To prove this, we consider a sequence of coverings of E_z by means of open arcs of $|z| = 1$ such that the total length of the arcs approaches $m(E_z)$, i.e. if $G_z^{(n)}$ denotes the set of all points in the nth covering, then

$$\lim_{n \to \infty} m(G_z^{(n)}) = m(E_z).$$

We have, of course,

$$E_z \subseteq \bigcap_n G_z^{(n)} = G_z \quad \text{and} \quad m(G_z - E_z) = 0.$$

If G_w denotes the image of G_z on C, then $E_w \subset G_w$ and $G_w - E_w$ is measurable with $m(G_w - E_w) = 0$. Since G_w is measurable, being a set of type G_δ, it follows that E_w is measurable. That $m(E_w) > 0$ now follows from Theorem 3.3. The converse, that a set E_w of positive measure on C comes from a set E_z of positive measure on $|z| = 1$, is proved in a similar fashion.

4. Measure of boundary sets under conformal enlargement of domain

We prove next a theorem which allows us to compare the measures of two sets under a pair of conformal mappings; this theorem, which will be adequate for our purposes, is a special case of the elegant theory of the 'principle of enlargement of domain' of Nevanlinna ([1], pp. 63 ff.).

THEOREM 3.4. *Let a Jordan domain D_1 be contained in a second Jordan domain D_2, and let the intersection E of their frontiers be non-empty. Let $\zeta = f_1(z)$ and $\zeta = f_2(z)$ map D_1 and D_2 respectively onto $|\zeta| < 1$ in such a way that $f_1(z_0) = f_2(z_0) = 0$, where z_0 is some point in D_1. If E_1 and E_2 are the images of E on $|\zeta| = 1$ under the boundary extensions of $f_1(z)$ and $f_2(z)$ respectively, then*

$$m(E_1) \leqslant m(E_2).$$

We remark first that E is closed, so that E_1 and E_2 are closed and therefore measurable. The function $\zeta = f_2(z)$ maps D_2 onto $|\zeta| < 1$ and carries D_1 onto a domain D in $|\zeta| < 1$ whose frontier intersects $|\zeta| = 1$ in the set E_2; the point z_0 is mapped into $\zeta = 0$, which lies in D. Let $w = \varphi(\zeta)$, $\varphi(0) = 0$, map D conformally onto $|w| < 1$; the set E_2 is mapped onto a set E_w on $|w| = 1$ with the property that $m(E_1) = m(E_w)$. To prove Theorem 3.4 it suffices then to show that $m(E_w) \leqslant m(E_2)$.

We show first that $m(E_2) = 0$ implies that $m(E_w) = 0$, so that we may assume in the proof that $m(E_2) > 0$. Now if $m(E_2) = 0$, it follows from Lemma 1, Chapter 2, that there exists a function

$h(\zeta)$ which is analytic and bounded in $|\zeta| < 1$ and which has no radial limit at any point of E_2. The function $h(\varphi^{-1}(w))$ is analytic and bounded in $|w| < 1$ and, by the nature of $h(\zeta)$, cannot have a radial limit at any point of E_w. Hence it follows from Fatou's theorem that $m(E_w) = 0$, so that we may assume in what follows that $m(E_2) > 0$. We remark that a similar procedure shows that the image on $|w| = 1$ of any subset of E_2 of measure zero must also be of measure zero.

We define the function

$$u(\zeta) = \frac{1}{2\pi} \int_{E_2} \frac{1-r^2}{1+r^2-2r\cos(\theta-t)} dt \quad (\zeta = r\,e^{i\theta}),$$

which, by Corollary 1 in Chapter 2, is harmonic and bounded in $|\zeta| < 1$ and has radial limit 1 almost everywhere in E_2 and radial limit 0 almost everywhere in $\mathscr{C}E_2$. In addition, $u(\zeta) > 0$ on that part of $\mathrm{Fr}\,D$ lying inside $|\zeta| < 1$. Now if $v(\zeta)$ denotes a harmonic conjugate of $u(\zeta)$, the function $g(\zeta) = e^{u+iv}$ is analytic and bounded in $|\zeta| < 1$ and has radial limits of modulus e almost everywhere in E_2. The function $G(w) = g(\varphi^{-1}(w))$ is analytic and bounded in $|w| < 1$, and must therefore possess radial limits almost everywhere on $|w| = 1$. We assert next that $G(w)$ possesses radial limits of modulus e almost everywhere in E_w. Suppose that this were not the case and that there were a subset $E_w' \subset E_w$ with $m(E_w') > 0$ such that $G(w)$ had radial limits of modulus less than e on E_w'. Since the conformal mapping $\zeta = \varphi^{-1}(w)$ is analytic and bounded in $|w| < 1$, it follows from Fatou's theorem that almost all points of E_w' are mapped into points of E_2 which are accessible from D, and that almost all radii terminating in E_w' are mapped into curves lying in D and terminating at a set E_2' of accessible points of $E_2 \cap \mathrm{Fr}\,D$. Along such curves $g(\zeta)$ must tend to limits of modulus less than e. By Lindelöf's theorem the radial limits of $g(\zeta)$ at E_2' must exist and have modulus less than e; this implies that $m(E_2') = 0$. From our remarks above it follows that $m(E_w') = 0$, so that $G(w)$ must possess radial limits of modulus e almost everywhere in E_w. We now define the function

$$u_1(w) = \frac{1}{2\pi} \int_{E_w} \frac{1-\rho^2}{1+\rho^2-2\rho\cos(\psi-t)} dt \quad (w = \rho\,e^{i\psi}),$$

which is harmonic and bounded in $|w| < 1$, and which has radial limits 1 almost everywhere in E_w and 0 almost everywhere in $\mathscr{C}E_w$. If $v_1(w)$ is a harmonic conjugate of $u_1(w)$, the function $G_1(w) = e^{u_1+iv_1}$ is analytic in $|w| < 1$ with $1 < |G_1(w)| < e$, and the function $F(w) = G(w)/G_1(w)$ is analytic and bounded in $|w| < 1$. Almost everywhere in E_w, $|F(w)|$ has radial limit 1, whereas almost everywhere in $\mathscr{C}E_w$, $|F(w)|$ has radial limit greater than 1 since $G_1(w)$ has radial limit 1 and $|G(w)|$ has radial limit greater than 1 almost everywhere in $\mathscr{C}E_w$. Hence

$$1 < |F(0)| = e^{u(0)-u_1(0)},$$

which implies that

$$u(0) - u_1(0) = m(E_2) - m(E_w) \geqslant 0,$$

so that the theorem is proved.

Theorem 3.4 is a particular case of a quite general principle,[†] as we have already pointed out; we remark that the proof, as given here, is applicable to the case where D_1 and D_2 are arbitrary simply connected domains rather than Jordan domains.

5. Angular limits of univalent functions

The extension of Fatou's theorem to the class of univalent (or schlicht) functions in $|z| < 1$ is quite simple and consists of nothing more than the remark that, under the succession of elementary mappings by which the mapping of $|z| < 1$ onto the general simply connected domain is reduced to the mapping of $|z| < 1$ onto a *bounded* simply connected domain, sets of measure zero are preserved. The hypothesis of univalence, however, is highly restrictive, and it is not surprising that the set of points on $|z| = 1$ where $f(e^{i\theta})$ does not exist can be shown to form a much thinner set for univalent functions than for general bounded functions. We prove next a theorem of Beurling [1][‡] which strengthens the Fatou and Riesz theorems in this way.

THEOREM 3.5. *If $w = f(z)$ is analytic and univalent in $|z| < 1$, then, for all $e^{i\theta}$ except for a possible set of capacity zero on $|z| = 1$,*
$$\lim_{z \to e^{i\theta}} f(z) \text{ exists uniformly inside any angular region in } |z| < 1 \text{ with}$$

† Nevanlinna [1].
‡ The proof given here is that of Tsuji [2].

vertex at $e^{i\theta}$. Furthermore, $f(e^{i\theta}) = \lim\limits_{r \to 1} f(r\,e^{i\theta})$ cannot be constant on any set of positive capacity on $|z| = 1$.

Because of the nature of the functions which reduce the problem to that of the bounded univalent function, we remark first that there is no loss of generality if we assume throughout that $|f(z)| < M$ in $|z| < 1$. Now let $e^{i\theta}$ be an arbitrary point of $|z| = 1$, and let $l_t(-\tfrac{1}{2}\pi < t < \tfrac{1}{2}\pi)$ be a rectilinear segment of length $\cos t$ lying in $|z| < 1$ and forming an angle t at $e^{i\theta}$ with the radius drawn to $e^{i\theta}$. We form the integral

$$L(\theta, t) = \int_{l_t} |f'(z)|\,|dz|, \tag{3.5.1}$$

which represents the length, whenever it exists, of the image of l_t under the mapping $w = f(z)$. Clearly, if (3.5.1) is finite, then $\lim f(z)$ must exist as $z \to e^{i\theta}$ along l_t.

We consider the integral

$$\Lambda(\theta) = \int_{-\frac{1}{2}\pi}^{\frac{1}{2}\pi} L(\theta, t)\ \cos t\,dt. \tag{3.5.2}$$

It is clear that, whenever $\Lambda(\theta)$ is finite, then $L(\theta, t)$ is finite for all t in $-\tfrac{1}{2}\pi < t < \tfrac{1}{2}\pi$ except for a set of linear measure zero. Our method of proof will be to show that $\Lambda(\theta)$ is finite for all $e^{i\theta}$ on $|z| = 1$ except for a possible set of capacity zero. Thus, for any $e^{i\theta}$ not belonging to a certain set of capacity zero, $L(\theta, t)$ will be finite for almost all t in $-\tfrac{1}{2}\pi < t < \tfrac{1}{2}\pi$, so that, for all these values of t for which $L(\theta, t)$ is finite, the limit of $f(z)$ must exist as $z \to e^{i\theta}$ along the rectilinear segment making an angle t with the radius drawn to $e^{i\theta}$. Since $|f(z)| < M$ in $|z| < 1$, it then follows from Lindelöf's theorem that $f(z)$ tends uniformly to a unique limit as $z \to e^{i\theta}$ along a non-tangential path. In order to complete the first part of Theorem 3.5, we must then prove that $\Lambda(\theta)$ is finite for all $e^{i\theta}$ on $|z| = 1$ with the possible exception of a set of capacity zero.

Let us suppose, to the contrary, that $\Lambda(\theta)$ is infinite on a set of positive capacity. By definition, such a set must contain a closed subset E of positive capacity. There exists then a distribution $\mu(\xi)$ of a total mass equal to 1 on E such that the potential

$$U(z) = \int_E \log\frac{1}{|z-\xi|}\,d\mu(\xi)$$

is bounded, $|U(z)| < V_0 < \infty$, for $|z| < \infty$. In particular

$$U(z) = U(r, \theta), \quad r < 1,$$

is harmonic in $|z| < 1$ and, since the complement of E carries no mass, may be written in the form

$$U(r, \theta) = \int_0^{2\pi} \log \frac{1}{|e^{i\varphi} - r\, e^{i\theta}|} d\mu(\varphi), \qquad (3.5.3)$$

where we have written $\mu(\varphi)$ for

$$\int_{\xi \in E_\varphi} d\mu(\xi) = \int_{t=0}^{\varphi} d\mu(e^{it}),$$

E_φ being the set $E \cap A_\varphi$ and A_φ the arc of $|z| = 1$ defined by $z = 1$ and $z = e^{i\varphi}$, for φ increasing.

The function $\mu(\varphi)$ is an additive function of sets on the closed interval $[0, 2\pi]$ and has a Fourier–Stieltjes series (Zygmund [2], I, p. 11)

$$d\mu(\varphi) \sim \sum_{n=1}^{\infty} (a_n \cos n\varphi + b_n \sin n\varphi), \qquad (3.5.4)$$

where $\quad a_n = \dfrac{1}{\pi} \displaystyle\int_0^{2\pi} \cos nt\, d\mu(t), \quad b_n = \dfrac{1}{\pi} \displaystyle\int_0^{2\pi} \sin nt\, d\mu(t).$

Further,

$$\log \frac{1}{|e^{i\varphi} - r\, e^{i\theta}|} = \tfrac{1}{2} \log \frac{1}{1 - 2r \cos(\theta - \varphi) + r^2}$$

$$= \sum_{\nu=1}^{\infty} \frac{\cos \nu(\theta - \varphi)}{\nu} r^\nu$$

$$= \sum_{\nu=1}^{\infty} \frac{r^\nu}{\nu} (\cos \nu\theta \cos \nu\varphi + \sin \nu\theta \sin \nu\varphi). \quad (3.5.5)$$

Applying Parseval's theorem (Zygmund [2], I, p. 157) to (3.5.3) with the expressions (3.5.4) and (3.5.5) under the sign of integration we obtain the trigonometrical series expression

$$U(r, \theta) = \sum_{n=1}^{\infty} \frac{r^n}{n} (a_n \cos n\theta + b_n \sin n\theta).$$

A further application of Parseval's theorem now gives

$$\int_0^{2\pi} U(r, \theta)\, d\mu(\theta) = \sum_{n=1}^{\infty} \frac{1}{n} (a_n^2 + b_n^2),$$

whence
$$\sum_{n=1}^{\infty} \frac{1}{n}(a_n^2 + b_n^2) \leqslant V_0. \qquad (3.5.6)$$

Thus it follows that

$$\iint_{|z|<1} \left(\frac{\partial U}{\partial r}\right)^2 r\, dr\, d\theta = \frac{\pi}{2} \sum_{n=1}^{\infty} \frac{1}{n}(a_n^2 + b_n^2) \leqslant \frac{\pi}{2} V_0. \qquad (3.5.7)$$

If we define the area integral

$$S = \iint_{|z|<1} |f'(z)| \frac{\partial U}{\partial r} r\, dr\, d\theta, \qquad (3.5.8)$$

it follows from Schwarz's inequality that

$$S^2 \leqslant \iint_{|z|<1} |f'(z)|^2 r\, dr\, d\theta \iint_{|z|<1} \left(\frac{\partial U}{\partial r}\right)^2 r\, dr\, d\theta < \pi M^2 \tfrac{1}{2}\pi V_0 < \infty, \qquad (3.5.9)$$

since the image domain of $|z| < 1$ under $w = f(z)$ is contained in the circle $|w| \leqslant M$. Because $U(r, \theta)$ is harmonic, we may write

$$\frac{\partial U}{\partial r} r\, d\theta = -\int_E d\arg(r\, e^{i\theta} - \zeta)\, d\mu(\zeta),$$

so that $S = -\displaystyle\int_{\zeta \in E} d\mu(\zeta) \iint_{|z|<1} |f'(z)|\, d\arg(r\, e^{i\theta} - \zeta)\, dr.$

If we set $\sigma(\zeta) = -\displaystyle\iint_{|z|<1} |f'(z)|\, d\arg(r\, e^{i\theta} - \zeta)\, dr,$

we have
$$S = \int_{\zeta \in E} \sigma(\zeta)\, d\mu(\zeta). \qquad (3.5.10)$$

We shall show that $\sigma(\zeta) = \infty$ for each point $\zeta \in E$; from the fact that S in (3.5.9) and (3.5.10) is finite, it will follow that the assumption that E has positive capacity is untenable. We may assume for simplicity that $\zeta = 1$; we have that

$$t = -\arg(r\, e^{i\theta} - 1), \quad \frac{-\pi}{2} < t < \frac{\pi}{2}, \qquad (3.5.11)$$

so that $\dfrac{dt}{d\theta} = \dfrac{r\cos\theta - r^2}{1 + r^2 - 2r\cos\theta}.$

We fix $r = \cos\theta_0$, so that the point $z = r\, e^{i\theta}$ lies outside the circle $r = \cos\theta$ whenever $\theta_0 \leqslant |\theta| \leqslant \pi$, and we observe that $dt \geqslant 0$ for

$|\theta| \leqslant \theta_0$ and $dt \leqslant 0$ for $\theta_0 \leqslant |\theta| \leqslant \pi$. Now, for $\theta_0 \leqslant |\theta| \leqslant \pi$, we have that

$$0 \leqslant \frac{r - \cos\theta}{1 + r^2 - 2r\cos\theta} \leqslant \frac{1}{1+r} \leqslant 1,$$

so that $|dt| \leqslant r\,d\theta$. From these estimates, it follows that

$$\int_0^1 dr \int_{\theta_0 \leqslant |\theta| \leqslant \pi} |f'(z)|\,|dt| \leqslant \iint_{|z| \leqslant 1} |f'(z)|\,r\,dr\,d\theta$$

$$\leqslant \left(\iint_{|z|<1} |f'(z)|^2\,r\,dr\,d\theta \right)^{\frac{1}{2}} \left(\iint_{|z|<1} r\,dr\,d\theta \right)^{\frac{1}{2}}$$

$$\leqslant \sqrt{(\pi M^2)}\,\sqrt{(\pi)} = \pi M.$$

Now, for $\zeta = 1$, we have

$$\sigma(1) = \iint_{|z|<1} |f'(z)|\,dt\,dr,$$

so that

$$\sigma(1) \geqslant \int_0^1 dr \int_{|\theta| \leqslant \theta_0} |f'(z)|\,dt - \pi M = \iint_{|z-\frac{1}{2}| \leqslant \frac{1}{2}} |f'(z)|\,dr\,dt - \pi M. \tag{3.5.12}$$

If, in this last integral, we set

$$\rho\,e^{it} = 1 - r\,e^{-i\theta}, \quad dr\,dt = \frac{\cos t - \rho}{1 + \rho^2 - 2\rho\cos t}\,d\rho\,dt,$$

and observe that $1 + \rho^2 - 2\rho\cos t \leqslant 1$ in $|z - \frac{1}{2}| \leqslant \frac{1}{2}$, we have that

$$\iint_{|z-\frac{1}{2}| \leqslant \frac{1}{2}} |f'(z)|\,dr\,dt = \int_{-\frac{1}{2}\pi}^{\frac{1}{2}\pi} dt \int_0^{\cos t} \frac{|f'(z)|\,(\cos t - \rho)}{[1 + \rho^2 - 2\rho\cos t]^{\frac{1}{2}}}\,d\rho$$

$$\geqslant \int_{-\frac{1}{2}\pi}^{\frac{1}{2}\pi} dt \int_0^{\cos t} |f'(z)|\,(\cos t - \rho)\,d\rho.$$

If $0 \leqslant \rho \leqslant \frac{1}{2}\cos t$, then $\cos t - \rho \geqslant \frac{1}{2}\cos t$, so that the last integral is greater than

$$\int_{-\frac{1}{2}\pi}^{\frac{1}{2}\pi} \frac{\cos t}{2}\,dt \int_0^{\frac{1}{2}\cos t} |f'(z)|\,d\rho = \frac{1}{2} \int_{-\frac{1}{2}\pi}^{\frac{1}{2}\pi} L(0,t)\cos t\,dt = \frac{1}{2}\Lambda(0).$$

But $\Lambda(0)$ was assumed to be infinite for $\theta = 0$; this contradiction shows that E cannot have positive capacity, so that the first part of the theorem is proved.

In order to prove the second part of Theorem 3.5, we remark

first that the details of the above proof carry over almost without change if we use the spherical metric

$$ds = \frac{|f'(z)|\,|dz|}{1+|f(z)|^2},$$

instead of $|f'(z)|\,|dz|$, and if we assume that the area of the image of $|z| < 1$ under $f(z)$ on the Riemann sphere is bounded. In the previous proof we could make $f(z)$ bounded by a linear transformation and apply Theorem 2.3. This is not available to us here but, since the image under $f(z)$ of the unit disc is a Riemann surface of bounded area over the sphere, $f(z)$ takes almost all values only a finite number of times and instead of Theorem 2.3 we can apply Theorem 2.20 to show that the asymptotic values are Fatou values, i.e. uniformly approached in every Stolz angle. For we can choose three values which are omitted by $f(z)$ in an annulus whose outer boundary is the unit circle $|z| = 1$ and a conformal mapping of any sector of this onto a disc $|\xi| \leqslant 1$ brings us back to the conditions of Theorem 2.20. The first part of Theorem 3.5 may therefore be stated in the following more general form.

THEOREM 3.5.1. *If $f(z)$ is meromorphic in $|z| < 1$ with*

$$\iint_{|z|<1} \frac{|f'(z)|^2}{(1+|f(z)|^2)^2} r\,dr\,d\theta < \infty, \tag{3.5.13}$$

then, for all $e^{i\theta}$ on $|z| = 1$, except for a possible set of capacity zero, $\lim_{z \to e^{i\theta}} f(z)$ *exists uniformly inside every angular region with vertex at $e^{i\theta}$.*

We now have to prove the second part of Theorem 3.5. Although we cannot prove the full result under the hypotheses of Theorem 3.5.1, we shall start from these hypotheses and prove that if α is an *ordinary value*, according to the definition we shall give in a moment, and if E_α is the set in which $f(e^{i\theta}) = \alpha$, then cap $E_\alpha = 0$. For a univalent $f(z)$, and more generally for a (multivalent) function which takes no value more than N times, $N < \infty$, all values are ordinary and the second part of Theorem 3.5 follows.

Let $n(\alpha)$ be the number of α-points (zeros of $f(z) - \alpha$ if $\alpha \neq \infty$ or poles if $\alpha = \infty$) in $|z| < 1$. α is a point of the Riemann sphere and

the portion of the Riemann surface of $f(z)$ covering the spherical disc $\delta_\rho(\alpha)$ of centre α and radius ρ is mapped by the inverse of $f(z)$ onto a set (finite or countable) of domains in $|z| < 1$ the totality of which we denote by $\Delta_\rho(\alpha)$. Evidently

$$A_\rho(\alpha) = \iint_{\Delta_\rho(\alpha)} \frac{|f'(z)|^2}{(1 + |f(z)|^2)^2} r\, dr\, d\theta = \iint_{\delta_\rho(\alpha)} n(w)\, d\sigma,$$

where $d\sigma$ is the spherical surface element.

If
$$\bar{n}(\alpha) = \limsup_{\rho \to 0} \frac{A_\rho(\alpha)}{\pi\rho^2}$$

we say that α is an *ordinary value* in the sense of Beurling.

Since by hypothesis the total area of the surface is finite, $n(w)$ is finite for almost all w. Moreover, the surface integral of $n(w)$ is an additive function of sets and thus by a theorem of Lebesgue (Saks *Theory of the Integral*, p. 115) has a symmetrical derivative almost everywhere, that is to say

$$\lim_{\rho \to 0} \frac{A_\rho(w)}{\pi\rho^2} < \infty$$

exists almost everywhere. Therefore almost all values of w are ordinary values. For any value of w, $n(w) \leqslant \bar{n}(w)$, while for an ordinary value $n(w) = \bar{n}(w)$.

Without loss of generality we take $\alpha = 0$ and ordinary. Denoting $\Delta_\rho(0)$ by Δ_ρ we have accordingly

$$\iint_{\Delta_\rho} \frac{|f'(z)|^2}{(1 + |f(z)|^2)^2} r\, dr\, d\theta < K\rho^2 \tag{3.5.14}$$

for some constant K and sufficiently small ρ. We now assume $E = E_0$ to be of positive capacity and, as before, we consider the bounded potential $U(z)$ defined by (3.5.3) for the set E, together with the integral

$$S_\rho = \iint_{\Delta_\rho} \frac{|f'(z)|}{1 + |f(z)|^2} \frac{\partial U}{\partial r} r\, dr\, d\theta. \tag{3.5.15}$$

As in the proof of the first part of Theorem 3.5, we have

$$S_\rho^2 \leqslant K\rho^2 \iint_{\Delta_\rho} \left(\frac{\partial U}{\partial r}\right)^2 r\, dr\, d\theta \leqslant K\rho^2 \iint_{|z|<1} \left(\frac{\partial U}{\partial r}\right)^2 r\, dr\, d\theta.$$

Now, since
$$\iint_{|z|<1} \left(\frac{\partial U}{\partial r}\right)^2 r\,dr\,d\theta < \infty,$$

we have
$$\lim_{\rho\to 0} \iint_{\Delta_\rho} \left(\frac{\partial U}{\partial r}\right)^2 r\,dr\,d\theta = 0,$$

and hence
$$\lim_{\rho\to 0} \frac{S_\rho}{\rho} = 0. \tag{3.5.16}$$

Writing
$$\sigma_\rho(\zeta) = -\iint_{\Delta_\rho} \frac{|f'(z)|}{1+|f(z)|^2}\, d\arg\,(r\,e^{i\theta}-\zeta)\,dr,$$

we have
$$S_\rho = \int_{\zeta\in E} \sigma_\rho(\zeta)\,d\mu(\zeta). \tag{3.5.17}$$

We suppose now that $\zeta = 1$ belongs to E and that $\lim_{r\to 1} f(r) = 0$; we have also that $\lim_{z\to 1} f(z) = 0$ in the angular domain
$$|\arg\,(z-1)| \leqslant \tfrac{1}{4}\pi.$$

Let Δ denote the common part of Δ_ρ and this angular domain; then, for ρ sufficiently small, Δ lies inside $|z-\tfrac{1}{2}| \leqslant \tfrac{1}{2}$, so that, by analogy with (3.5.12), we have

$$\begin{aligned}
\sigma_\rho(1) &\geqslant \iint_\Delta \frac{|f'(z)|}{1+|f(z)|^2}\,dr\,dt \\
&\quad - \left\{ \iint_{\Delta_\rho} \left(\frac{|f'(z)|}{1+|f(z)|^2}\right)^2 r\,dr\,d\theta \iint_{\Delta_\rho} \left(\frac{\partial U}{\partial r}\right)^2 r\,dr\,d\theta \right\}^{\frac{1}{2}} \\
&\geqslant \iint_\Delta \frac{|f'(z)|}{1+|f(z)|^2}\,dr\,dt - \epsilon_1\rho \\
&\geqslant \int_{-\frac{1}{4}\pi}^{\frac{1}{4}\pi} \frac{\cos t}{2}\,dt \int_{l_t} \frac{|f'(z)|}{1+|f(z)|^2}\,|dz| - \epsilon_1\rho,
\end{aligned}$$

where l_t is a rectilinear segment of Δ terminating at $z = 1$, and where ϵ_1 is a real number such that $\lim_{\rho\to 0} \epsilon_1 = 0$. Since l_t joins $z = 1$ and a boundary point of Δ_ρ on which
$$\frac{|w|}{\sqrt{(1+|w|^2)}} = \rho,$$

the length of the image of l_t under $w = f(z)$ is not less than ρ, so

that $\sigma_\rho(1) \geqslant \rho/\sqrt{2} - \epsilon_1\rho$. Similarly for each $\zeta \in E$, so that (3.5.17) implies that

$$S_\rho \geqslant \frac{\rho}{\sqrt{2}} - \epsilon_1\rho, \quad \text{or} \quad \frac{1}{\sqrt{2}} - \epsilon_1 \leqslant \frac{S_\rho}{\rho},$$

which contradicts (3.5.16). This contradiction shows that the set where $f(e^{i\theta}) = 0$ cannot have positive capacity, and the second part of Theorem 3.5 is proved.

6. Correspondence of sets of capacity zero under conformal mapping

The theorem of Beurling, which was proved in § 5, has had wide application, particularly in connexion with the sets of convergence of various classes of trigonometrical series, but we are interested in its application to the theory of conformal mapping. In § 3 we pointed out that it is possible for a set of positive linear measure (in the sense defined there) on a Jordan curve C to be carried into a set of measure zero on $|z| = 1$ under the conformal mapping onto $|z| < 1$ of the Jordan domain determined by C; such a curve C cannot be rectifiable, of course, by Theorem 3.3. Beurling's theorem enables us to prove a result† in which the question of rectifiability does not enter.

THEOREM 3.6. *Let G be a simply connected domain in the w-plane bounded by a Jordan curve C, and let E_w be a set on C of capacity zero. Then the set E_z on $|z| = 1$ corresponding to E_w under a conformal mapping of G onto $|z| < 1$ is of capacity zero.*

It is clear that it suffices to assume E_w to be closed, for, if the theorem were false, then E_z must contain a closed subset of positive capacity, so that a closed subset of E_w maps into a closed subset of E_z having positive capacity. Assuming E_w to be closed, it follows from § 7 of Chapter 1 that we can find a distribution $d\mu$ of unit mass on E_w of such a nature that the potential

$$u(w) = \int_{\zeta \in E_w} \log \frac{1}{|w - \zeta|} d\mu(\zeta) \tag{3.6.1}$$

tends to $+\infty$ as $w \to \zeta \in E_w$. The level curves λ_ρ on which

† Tsuji [2].

$u(w) = \rho > 0$ consist of a finite number of simple closed curves enclosing the set E_w. If we set

$$u(w) + iv(w) = \int_{\zeta \in E_w} \log \frac{1}{w - \zeta} d\mu(\zeta), \qquad (3.6.2)$$

then
$$\int_{\lambda_\rho} dv = \int_{\lambda_\rho} \frac{\partial v}{\partial s} ds = \int_{\lambda_\rho} \frac{\partial u}{\partial n} ds = 2\pi, \qquad (3.6.3)$$

where ds denotes the element of arc length on λ_ρ, and where n is the inner normal for λ_ρ. The function

$$\xi = F(z) = u(f(z)) + iv(f(z))$$

is analytic in $|z| < 1$, and, since $u \to +\infty$ as $w \to \zeta \in E_w$, there is a positive constant k such that $u(w) + k \geqslant 1$ for any point w in G, so that the function $\eta = \eta(z) = [F(z) + k]^{\frac{1}{3}}$ is also analytic in $|z| < 1$. Since $d\eta = d\xi/[3(\xi + k)^{\frac{2}{3}}]$, the area A_η in the η-plane corresponding to the circle $|z| < 1$ under the function $\eta = [F(z) + k]^{\frac{1}{3}}$ is given by

$$A_\eta = \frac{1}{9} \iint_{R_\xi} \frac{du\, dv}{[(u+k)^2 + v^2]^{\frac{2}{3}}},$$

where the integral is taken over the Riemann configuration R_ξ of the function $\xi = u(w) + iv(w)$. Denoting by λ'_ρ that part of λ_ρ lying inside C, we have from (3.6.3) that

$$A_\eta \leqslant \frac{1}{9} \int_{1-k}^{\infty} du \int_{\lambda'_\rho} \frac{dv}{[(u+k)^2 + v^2]^{\frac{2}{3}}} \leqslant \frac{1}{9} \int_{1-k}^{\infty} \frac{du}{(u+k)^{\frac{4}{3}}} \int_{\lambda'_\rho} dv$$

$$\leqslant \frac{2\pi}{9} \int_{1-k}^{\infty} \frac{du}{(u+k)^{\frac{4}{3}}} < \infty.$$

Hence the function $\eta = \eta(z)$ fulfils the conditions of Theorem 3.5.1; since $\eta(z) \neq \infty$ in $|z| < 1$, it follows from our proof of the second part of Theorem 3.5 that the set of points on $|z| = 1$ for which $\lim_{r \to 1} \eta(r e^{i\theta}) = \infty$ is of capacity zero. Since we may write

$$[\eta(z)]^3 = \int_{\zeta \in E_w} \log \frac{1}{f(z) - \zeta} d\mu(\zeta) + k,$$

it follows that E_z must be of capacity zero.

CHAPTER 4

INTRINSIC PROPERTIES OF CLUSTER SETS

1. Global cluster set, range and asymptotic set

We have seen in Chapter 1 that, whenever a set E on the frontier of a domain D is of positive linear measure, the cluster set $C(f, z_0)$ at a point $z_0 \in E$ of a function $f(z)$, meromorphic in D, may be subtotal and non-degenerate, and we shall see later on that the general character of the theory in this situation emerges from a detailed study of the case where D is the unit circle $|z| < 1$. The definitions of $C(f, z_0)$, $R(f, z_0)$, and $\Gamma(f, z_0)$ given in Chapter 1 are equally applicable whether z_0 is a singularity or not; in the latter case, of course, $C(f, z_0)$ and $\Gamma(f, z_0)$ are degenerate and identical, while $R(f, z_0)$ is empty.

One important difference between the theory of the continuous and the discontinuous boundary must be emphasized at the outset. An asymptotic path of $f(z)$ in $|z| < 1$ may have as its 'end' on the circumference K: $|z| = 1$ not a point z_0 but an arc of K, or the whole of K, as is the case when the asymptotic path is a spiral. We must therefore consider cluster sets not only with respect to an individual point z_0 on K, that is, 'in the small', but also with respect to K itself (or to arcs of K), that is, 'in the large'. The theory of cluster sets in the large—which we may conveniently call *global* cluster sets—has a closer formal similarity to the theory in the case of an isolated point and is, in some respects, simpler than the theory in the small. We proceed now with the definitions of these new concepts and give some of their elementary properties.

(1) *A complex number α is said to belong to the (global) cluster set $C(f)$ of a function $f(z)$ defined in $|z| < 1$ if there exists a sequence $\{z_n\}$, $|z_n| < 1$, such that*

$$\lim_{n \to \infty} |z_n| = 1 \quad and \quad \lim_{n \to \infty} f(z_n) = \alpha.$$

It is easily seen that the following definition of $C(f)$ is equiva-

lent: If $A(n)$ denotes the annulus $1 - (1/n) < |z| < 1$ and $f(A(n))$ the image of $A(n)$ under the mapping $f(z)$, then $C(f) = \bigcap_n \overline{f(A(n))}$.

(2) *The value α is said to belong to the range $R(f)$ of $f(z)$ if there exists a sequence $\{z_n\}$, $|z_n| < 1$, such that*

$$\lim_{n \to \infty} |z_n| = 1 \quad and \quad f(z_n) = \alpha$$

for each n.

In terms of the sets $A(n)$ and $f(A(n))$ of (1) we have, alternatively, that $R(f) = \bigcap_n f(A(n))$.

(3) *A value α belongs to the asymptotic set $\Gamma(f)$ of a function $f(z)$ if there exists a continuous curve $z = z(t)$, $0 \leqslant t < 1$, in $|z| < 1$, such that*

$$\lim_{t \to 1} |z(t)| = 1 \quad and \quad \lim_{t \to 1} f(z(t)) = \alpha.$$

We show next some rather easy consequences of the definitions (1)–(3).

THEOREM 4.1. *If $f(z)$ is meromorphic in $|z| < 1$ and of unbounded characteristic, then $C(f)$ is total.*

For if there exists a point c not in $C(f)$, then $[f(z) - c]^{-1}$ is bounded in $|z| < 1$, and, since $N(r, c) = O(1)$, the assertion follows from (2.9.1). The converse is not necessarily true, however, as is shown by simple examples, e.g. the function given by (2.9.2).

Theorem 4.1 depends on the analytic character of $f(z)$, and, as a contrast, the next property requires only that a continuous function be a *light, interior mapping*,† that is, the set of points $f(A)$ shall be open whenever the set A in $|z| < 1$ is open, and that $f^{-1}(w_0)$ shall be totally disconnected for each $w_0 \in f(A)$.

THEOREM 4.2. *If $f(z)$ is a light, interior mapping defined in $|z| < 1$, then $R(f)$ is dense in the interior of $C(f)$. Also $\mathscr{C}R(f)$ is of Category I in the interior of $C(f)$.*‡

Each of the sets $f(A(n))$, where $A(n)$ is the annulus defined in (1) above, is dense in the interior of $C(f)$, so that the intersection $\bigcap_n f(A(n)) = R(f)$ must also be dense in the interior of $C(f)$. By

† See, for example, Whyburn [1] or [3].
‡ For the definition of first category (also called Category I), see p. 75.

definition, $R(f)$ is a G_δ and, being dense, is residual, i.e. the complement of a set of Category I, in the interior of $C(f)$. This proves the theorem.

As an interesting illustration of Theorem 4.2, we cite the example of a Blaschke product $B(z)$. If $B(z)$ has only finitely many factors, then $C(B)$ consists of the circumference $|w| = 1$, $R(B)$ is empty, and Theorem 4.2 is vacuous. Theorem 4.2 has meaning then with respect to $B(z)$ only when $B(z)$ has infinitely many factors, for it follows from Theorem 2.14 that $R(B)$ is the open disc $|w| < 1$ with the possible deletion of a set of capacity zero. Since $|B(z)| < 1$, Theorem 4.2 now implies that $C(B)$ is the closed disc $|w| \leqslant 1$. Trivially, we have the

REMARK. *Under the hypotheses of Theorem 4.2, a necessary and sufficient condition that $C(f)$ be total is that $R(f)$ be everywhere dense.*

2. Existence theorems for global cluster sets

The local cluster sets $C(f, z_0)$, $R(f, z_0)$ and $\Gamma(f, z_0)$ have been defined in Chapter 1; these sets may be obtained from the definitions (1)–(3) by the additional requirement that the sequence $\{z_n\}$ in (1) and (2) converge to $z_0 = e^{i\theta_0}$ and that the arc $z = z(t)$ in (3) have the property that $\lim_{t \to 1} z(t) = z_0 = e^{i\theta_0}$. We remark that analogues of Theorem 4.2 and the remark following it apply equally well to the sets $C(f, z_0)$ and $R(f, z_0)$; indeed, if the annulus $A(n)$ is replaced by that part of an open disc, with radius $1/n$ and centre z_0, lying in $|z| < 1$, the arguments are identical.

Although the sets $C(f)$ and $C(f, z_0)$ are in many ways quite similar in their superficial behaviour, the first two theorems which we shall prove emphasize a fundamental difference between the two cluster sets and give rise to a question, which has not yet been resolved, concerning *the characterization of those continua κ for which there exists a function $f(z)$, meromorphic in* $|z| < 1$, *such that $C(f) = \kappa$.* We shall prove in Theorem 4.3 that for every continuum κ in the extended plane (or on the Riemann sphere) and for any point $z_0 = e^{i\theta_0}$ on $|z| = 1$, there exists† a meromorphic function in $|z| < 1$ such that $C(f, z_0) = \kappa$. The distinction

† Gross [1].

between the sets $C(f, z_0)$ and $C(f)$ is brought into sharper focus by Theorem 4.4,† which shows that there exists a continuum \mathscr{K} which cannot be a global cluster set $C(f)$ for any meromorphic function $f(z)$. To show that this difference between $C(f)$ and $C(f, z_0)$ depends on the restrictive properties of the interiority of $f(z)$, we prove in Theorem 4.5 that there exists in $|z| < 1$ a continuous function $f(z)$ such that $C(f)$ can be made to coincide with any given continuum κ. In contrast with this, an example of a continuum κ which cannot be the global cluster set of any light interior mapping of $|z| < 1$ has been given recently by Church [2].

THEOREM 4.3. *For every continuum κ on the w-sphere and an arbitrary point z_0 on $|z| = 1$, there exists a function $w = f(z)$, meromorphic in $|z| < 1$, such that $C(f, z_0) = \kappa$.*

We may assume that $z_0 = 1$ and that κ is not the entire plane, for this case is handled by the function $f(z) = \exp\{(z-1)^{-3}\}$, for which $C(f, 1)$ is total. The case that κ consists of a single point is trivial, so that we may assume that κ is not degenerate, and by means of a linear fractional transformation if necessary, that κ is bounded.

Let κ be contained in a square A of side a which is then sub-divided into n^2 congruent smaller squares $A_{n,j}$, and let

$$H_n = A_{n,1} \cup A_{n,2} \cup ... \cup A_{n,k_n}$$

denote the union of all those squares $A_{n,j}$ containing points of κ in their interiors or on their sides. It is clear that the interior G_n of H_n is a connected set of finite connectivity with the property that no point of G_n is at a distance greater than $a\sqrt{2}/2^n$ from κ. From G_n we may construct a simply connected domain G_n^* by the insertion of canonical cuts, and we have

$$\bigcap_n \bar{G}_n^* = \bigcap_n \bar{G}_n = \kappa.$$

Now inside each G_n^* we can form a simply connected domain D_n in the form of a narrow ribbon which passes at least once through each of the squares $A_{n,j}(j = 1, ..., k)$. The D_n may be thought of as forming parts of sheets of a surface R lying over A: the joining of $A_{n,j}$ to $A_{n+1, i}$ on R may be achieved in such a way that R is simply connected and does not pass outside G_n^*. We may now

† Potyagaĭlo [1]; the same example was given independently by Rudin [1].

map R conformally onto an infinite strip R_ζ in the ζ-plane, $\operatorname{Re}\zeta > 0, 0 < \operatorname{Im}\zeta < 1$, in such a way that the sides of the D_n are mapped into segments of the horizontal lines bounding R_ζ. If R_ζ is mapped onto $|z| < 1$ in such a way that $\zeta = \infty$ is carried into $z = 1$, we obtain a function $w = f(z)$ which has the desired property that $C(f, 1) = \kappa$. This theorem of Gross has recently been sharpened by Church [3] who proved that, given κ, there exists a function f analytic and bounded in $|z| < 1$ such that $C_\lambda(f, 1) = \kappa$, where λ is any path in $|z| < 1$ terminating at $z = 1$.

We prove next the interesting theorem that not every continuum in the plane can be a cluster set $C(f)$. We consider the continuum \mathscr{K} (Fig. 4.1) consisting of the circle $|w| = 1$ together

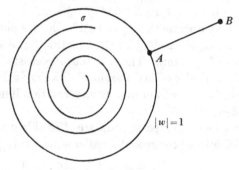

Fig. 4.1.

with a segment AB and a spiral σ tending asymptotically to $|w| = 1$. To show that there cannot exist a meromorphic function $f(z)$ with $C(f) = \mathscr{K}$, we assume otherwise; i.e. we suppose that there exists a simply connected Riemann surface R of hyperbolic type with the properties:

(1) Over every closed subdomain \bar{G} of \mathscr{CK}, R has at most a finite number of sheets.

(2) All boundary components—and only these—lie over \mathscr{K}.

First of all, we remark that it follows from (1) and from the form of \mathscr{K} that R has a finite number of sheets over the extended complex plane, so that R can have at most a finite number of algebraic branch points. Now if P is an arbitrary point of R lying over σ, we denote by σ_P that arc on R which is obtained by unrestricted continuation of f^{-1} from P along σ, and we consider the

set D_0 of those arcs σ_P which are homeomorphic to the closed segment [0, 1]. Thus a slit on all these arcs must decompose R, and from the fact that R has a finite number of sheets it follows that D_0 is finite. Consequently, for sufficiently small $\epsilon > 0$, that part of R over the annulus $A(\epsilon)$: $1 - \epsilon < |w| < 1$ contains no branch points, and an arbitrary or a regular arc of R, lying over the spiral $\sigma \cap A(\epsilon)$, coincides with it.

Now let $\gamma_1, \gamma_2, ..., \gamma_k$ denote all possible closed contours consisting of interior points of R and lying over the circle $|w| = \rho$, where $1 - \epsilon' < \rho < 1$, $\epsilon' < \epsilon$. From what we have proved above, it follows that, by means of the sections $\gamma_1, \gamma_2, ..., \gamma_k$, it is possible to single out from R a region $R^* \subset R$ situated precisely over the circle $|w| < 1$ and containing the boundary points of R over σ. Since R^* cannot contain any point of R whose modulus is greater than 1, the surface R cannot contain boundary points lying over the interval AB. This fact contradicts our assumption that R is simply connected, and we have†

THEOREM 4.4. *There exists a continuum \mathscr{K} in the w-plane which is not the global cluster set of any function meromorphic in $|z| < 1$.*

We show next that, in order to obtain a result analogous to Theorem 4.3 for $C(f)$ instead of $C(f, z_0)$, we must go to the class of functions which are continuous in $|z| < 1$; the simplified construction in the following theorem is due to G. S. Young.

THEOREM 4.5. *For any continuum κ in the w-plane there is a function $f(z)$, continuous in $|z| < 1$, for which $C(f) = \kappa$.*

We use the notation of the proof of Theorem 4.3. Let I_n denote the closed interval

$$\frac{n}{n+1} \leqslant r \leqslant \frac{n+1}{n+2} \quad (n = 0, 1, 2, ...),$$

so that $\bigcup_n I_n$ is the interval $0 \leqslant r \leqslant 1$, and let $\varphi_n(r)$ be a continuous mapping of I_n onto a curve lying inside G_n and passing at least once through each of the squares $A_{n,j}$ used in the definition of G_n. By defining the functions $\varphi_n(r)$ so that

$$\varphi_n\left(\frac{n+1}{n+2}\right) = \varphi_{n+1}\left(\frac{n+1}{n+2}\right),$$

† The above proof is due to Potyagaĭlo [1].

we may define a function $\varphi(r)$ on the interval $0 \leqslant r < 1$ by the rule

$$\varphi(r) = \varphi_n(r) \quad \text{for} \quad r \in I_n \quad (n = 0, 1, 2, \ldots).$$

The function $f(z) = \varphi(|z|)$ is then continuous in $|z| < 1$ and has the properties described in the statement of the theorem.

Before leaving this topic we point out that there are important classes of continua for which Theorem 4.5 holds when $f(z)$ is meromorphic in $|z| < 1$. For example, if κ is the boundary of a simply connected domain on which $f(z)$ maps $|z| < 1$ conformally then $C(f) = \kappa$.

3. Sets of maximum indetermination. Existence theorem for an arbitrary function

If \mathscr{G} is a closed, connected set lying in $|z| < 1$ except for one point z_0 on $|z| = 1$, we say that a value α lies in the cluster set $C_{\mathscr{G}}(f, z_0)$ if the sequence $\{z_n\}$ used in defining $C(f, z_0)$ is further restricted to lie in \mathscr{G}. The set $C_{\mathscr{G}}(f, z_0)$ is, of course, a closed subset of $C(f, z_0)$, and if, in addition, $f(z)$ is continuous in $|z| < 1$, then $C_{\mathscr{G}}(f, z_0)$ is also connected. Various choices of \mathscr{G} have suggested themselves naturally in the course of the classical developments in Chapter 2; for example, if \mathscr{G} is a radius of $|z| < 1$ drawn to the point $e^{i\theta}$, we designate the resulting cluster set by $C_\rho(f, e^{i\theta})$ and call it *the radial cluster set of $f(z)$ at the point $e^{i\theta}$*. In terms of the sets $C_\rho(f, e^{i\theta})$, Theorem 2.1 of Chapter 2 may be stated:

If $f(z)$ is analytic and bounded in $|z| < 1$, then $C_\rho(f, e^{i\theta})$ is degenerate for almost all $e^{i\theta}$ on $|z| = 1$.

More generally, if \mathscr{G} is an arc γ in $|z| < 1$ terminating at a point $e^{i\theta}$, the resulting cluster set $C_\gamma(f, e^{i\theta})$ is called a *curvilinear cluster set* at $e^{i\theta}$; this concept arises naturally in connexion with Lindelöf's theorem (Theorem 2.3, Chapter 2), which may be stated as follows:

If $f(z)$ is analytic and bounded in $|z| < 1$ and if for some $e^{i\theta}$ and some arc γ terminating at $e^{i\theta}$, $C_\gamma(f, e^{i\theta}) = a$, then $C_\rho(f, e^{i\theta}) = a$.

In particular, if γ is a chord $\rho(\varphi)$ of the unit circle passing through $e^{i\theta}$ and making an angle φ, $-\frac{1}{2}\pi < \varphi < \frac{1}{2}\pi$, with the radius to $e^{i\theta}$, we speak of the *chordal cluster set* $C_{\rho(\varphi)}(f, e^{i\theta})$ of $f(z)$ at $e^{i\theta}$.

Similarly, if \mathscr{G} is an angular region Δ formed by two chords passing through $e^{i\theta}$, we may define the corresponding angular cluster set $C_\Delta(f, e^{i\theta})$. Clearly if Δ_1 and Δ_2 are two angular regions with $\Delta_1 \subseteq \Delta_2$, then $C_{\Delta_1}(f, e^{i\theta}) \subseteq C_{\Delta_2}(f, e^{i\theta})$. The *outer angular cluster set* of $f(z)$ at $e^{i\theta}$ is now defined as the union

$$C_{\mathscr{A}}(f, e^{i\theta}) = \bigcup_\Delta C_\Delta(f, e^{i\theta})$$

taken over all angular regions Δ between pairs of chords passing through $e^{i\theta}$. Fatou's theorem (Theorem 2.4, Chapter 2) then asserts that *if $f(z)$ is analytic and bounded in $|z| < 1$, the outer angular cluster set $C_{\mathscr{A}}(f, e^{i\theta})$ is degenerate for almost all points of $|z| = 1$.*

While it is a fact that all the special cluster sets $C_{\mathscr{G}}(f, e^{i\theta})$ taken over sets \mathscr{G} of the type described above are subsets of $C(f, e^{i\theta})$, we should like to examine in some detail the question of whether they are always proper subsets. The example $w = f(z) = e^{(z+1)/(z-1)}$ shows that $C_\rho(f, 1) = C_{\mathscr{A}}(f, 1) = \{0\}$, whereas $C_\gamma(f, 1)$ consists of a closed circumference concentric with $|w| = 1$ whenever γ is a circular arc internally tangent to $|z| = 1$ at $z = 1$. On the basis of Theorem 2.14, Chapter 2, as applied to $f(z)$, it is easy to see that $C(f, 1)$ is the closed disc $|w| \leqslant 1$. It is natural to ask whether, for a given function $f(z)$ and for each point $e^{i\theta}$, there exists a set M in $|z| < 1$ such that $C_M(f, e^{i\theta}) = C(f, e^{i\theta})$; such a set, if it exists, is called a set of *maximum indetermination* for $f(z)$ at $e^{i\theta}$. We prove next that this question can be answered in the affirmative with M in the form of a simple arc γ terminating at $e^{i\theta}$; what is significant about Theorem 4.6 is that the proof requires nothing more about $f(z)$ than that it be defined in the unit disc.† Such a function we shall call an arbitrary function.

THEOREM 4.6. *If $f(z)$ is an arbitrary function defined in $|z| < 1$, and if $e^{i\theta}$ is an arbitrary point of $|z| = 1$, there exists a simple arc γ, lying in $|z| < 1$ and terminating at $e^{i\theta}$, such that $C_\gamma(f, e^{i\theta}) = C(f, e^{i\theta})$.*

Let $T_1, T_2, ..., T_n, ...$ denote a sequence of triangulations of the sphere with the properties that T_n is a refinement of T_{n-1} and that

† Although the original theorem (Collingwood [1]) presupposes $f(z)$ meromorphic, it is seen at once that this hypothesis is not used in the proof; the set $C(f, e^{i\theta})$ need not, however, be connected if $f(z)$ is not continuous.

ϵ_n, the maximum of the spherical diameters of the triangles constituting T_n, tends to zero as $n \to \infty$, i.e. the mesh of T_n tends to zero as $n \to \infty$. Let us denote by $\tau_{n,1}, \tau_{n,2}, ..., \tau_{n,m(n)}$ those triangles in T_n which meet the cluster set $C(f, e^{i\theta})$. We may arrange these triangles in a sequence $\{\tau_k\}$, $k = 1, 2, ...$, by means of the ordering

$$\tau_{1,1}, \tau_{1,2}, ..., \tau_{1,m(1)}, \tau_{2,1}, ..., \tau_{2,m(2)}, ...,$$
where

$$\tau_1 = \tau_{1,1}; \quad \tau_2 = \tau_{1,2}; \quad ...; \quad \tau_{m(1)} = \tau_{1,m(1)}; \quad \tau_{m(1)+1} = \tau_{2,1}; \quad$$

For any number $\rho > 0$, we denote by V_ρ the intersection of $|z| < 1$ with the circle $|z - e^{i\theta}| < \rho$. Then, for $\rho = \rho_1 = 1$, there exists in V_{ρ_1} a point z_1 such that $f(z_1) \in \bar\tau_1$, for otherwise the intersection $\tau_1 \cap C(f, e^{i\theta})$ would be empty. For $\rho = \rho_2 < |z_1 - e^{i\theta}|$, there exists in V_{ρ_2} a point z_2 such that $f(z_2) \in \bar\tau_2$. In general, there exists in V_{ρ_n} a point z_n such that $f(z_n) \in \bar\tau_n$, where ρ_n is chosen in such a way that $\rho_n < |z_{n-1} - e^{i\theta}|$ and $\lim_{n\to\infty} \rho_n = 0$; such a point z_n may be found since $\tau_n \cap C(f, e^{i\theta})$ cannot be empty for any n. Consequently, every point of $C(f, e^{i\theta})$ is a limit point of some subsequence of the sequence $\{z_n\}$. If we construct a simple arc γ which joins the points of the sequence $\{z_n\}$ in order, the condition $\lim_{n\to\infty} \rho_n = 0$ implies that γ terminates at $e^{i\theta}$, and, since $z_n \in \gamma$, it follows that $C_\gamma(f, e^{i\theta}) \supseteq C(f, e^{i\theta})$. This inclusion, when coupled with the trivial inclusion $C_\gamma(f, e^{i\theta}) \subseteq C(f, e^{i\theta})$, completes the proof of Theorem 4.6.

We remark that the choice of the sequence $\{z_n\}$ in the proof of Theorem 4.6 can be made in many ways; for example, we might have chosen the $\{z_n\}$ in such a way that $|z_n| < |z_{n+1}|$, and so that the arc γ increases monotonically, i.e. so that γ intersects each circle $|z| = r$ exactly once. Furthermore, the method of proof of Theorem 4.6 is easily modified to the case of the global cluster set $C(f)$, and we may state

THEOREM 4.7. *If $f(z)$ is an arbitrary function defined in $|z| < 1$, then there exists in $|z| < 1$ a continuous curve $\gamma: z = z(t)$, $0 \leqslant t < 1$, with $\lim_{t\to 1} |z(t)| = 1$, such that $C_\gamma(f) = C(f)$.*

The curve γ, which may be a spiral, can again be chosen to be monotone, i.e. for $t_1 < t_2$, $|z(t_1)| < |z(t_2)|$.

4. Maximality theorem for a continuous function

We have seen in § 3 that for any given function $f(z)$, defined in $|z| < 1$, and for each point $e^{i\theta}$ of $|z| = 1$, there exists a simple path γ terminating at $e^{i\theta}$ such that $C_\gamma(f, e^{i\theta}) = C(f, e^{i\theta})$. We shall show that, if $f(z)$ is taken to be continuous in $|z| < 1$, a stronger result may be proved; in fact, for all $e^{i\theta}$ on $|z| = 1$, except for a set which is 'thin' in a sense to be defined, the curve $\gamma(\theta)$ may be taken† to be a radius, or, more generally, the rotation of a given curve (or set) terminating at $z = 1$.

We shall say that a set E on $|z| = 1$ is *of first category* if E is the union of a countable set of nowhere dense sets; a set which is not of first category is said to be *of second category*. A set E on $|z| = 1$ is said to be *residual* on $|z| = 1$ if the complement of E on $|z| = 1$ is of first category. Although a set of first category is 'thin' in the sense that it is made up of countably many nowhere dense sets, we remark that it can be large in a measure-theoretic sense. For example, we can construct a perfect, nowhere dense set E_1 on $I: 0 \leqslant x \leqslant 1$ much like the well-known Cantor 'middle-thirds' set except that $m(E_1) = \frac{1}{2}$. Now in each of the intervals which form the complement of E_1 with respect to I we insert a similar perfect nowhere dense set whose linear measure is half that of the interval; the union of these sets, together with E_1, forms a perfect nowhere dense set E_2 on $[0, 1]$ with $m(E_2) = \frac{3}{4}$. Similarly, we can construct the perfect nowhere dense set E_n on $[0, 1]$ with $m(E_n) = 1 - \dfrac{1}{2^n}$. The set $E = \bigcup_n E_n$ is a set of first category on $[0, 1]$ with the property that $m(E) = 1$. The complement $\mathscr{C}E$ of E on $[0, 1]$ is then a residual set with $m(\mathscr{C}E) = 0$. We remark that, while the complement of a set of first category is always of second category, the converse is not necessarily true.

Let \mathscr{G}_0 be a continuum in $|z| \leqslant 1$ such that the intersection of \mathscr{G}_0 with $K: |z| = 1$ is the point $z = 1$, and let \mathscr{G}_θ denote the set obtained from \mathscr{G}_0 by a rotation about the origin through an angle θ, so that $\mathscr{G}_\theta \cap K$ is the point $e^{i\theta}$. In view of Theorem 4.6, the

† Collingwood [3, 8].

important case is that in which \mathscr{G}_0 is a simple arc terminating at $z = 1$. With these preliminaries we now prove the following *maximality theorem* (Collingwood [8]):

THEOREM 4.8. *If the real or complex function $f(z)$ is continuous in $|z| < 1$, and if $\{\mathscr{G}_\theta\}$ is the family of rotations of a continuum \mathscr{G}_0 such that $\mathscr{G}_0 \cap K$ is the point $z = 1$, then $C_{\mathscr{G}_\theta}(f, e^{i\theta}) = C(f, e^{i\theta})$ on a residual set of points $e^{i\theta}$ on $K: |z| = 1$.*

We assume, to the contrary, that there is a set E of second category on $|z| = 1$ at which $C_{\mathscr{G}_\theta}(f, e^{i\theta}) \neq C(f, e^{i\theta})$.

In three steps we shall select a subset of E, also of second category and therefore dense on an arc β of κ, in which a certain closed subset T_M of the Riemann sphere S meets $C(f, e^{i\theta})$ while $f(z)$ is uniformly bounded away from T_M throughout an annular wedge abutting on an arc $\beta' \subset \beta$. These properties are inconsistent.

The selection is carried out as follows.

(i) For any given $\epsilon > 0$ we denote by $C_{\mathscr{G}_\theta}(f, e^{i\theta})_{+\epsilon}$ the closed set of points on the Riemann sphere whose (spherical) distance from $C_{\mathscr{G}_\theta}(f, e^{i\theta})$ does not exceed ϵ. Since $C_{\mathscr{G}_\theta}(f, e^{i\theta}) \subseteq C(f, e^{i\theta})$, and since $C_{\mathscr{G}_\theta}(f, e^{i\theta}) \neq C(f, e^{i\theta})$ for $e^{i\theta} \in E$, there is a positive ϵ for which $C(f, e^{i\theta}) \cap \mathscr{C}(C_{\mathscr{G}_\theta}(f, e^{i\theta})_{+\epsilon})$ is not empty. Now let us choose a decreasing sequence of positive numbers $\{\epsilon_n\}$ with $\lim \epsilon_n = 0$. For a given n, let E_n be the set of points $e^{i\theta}$ such that

$$C(f, e^{i\theta}) \cap \mathscr{C}(C_{\mathscr{G}_\theta}(f, e^{i\theta})_{+\epsilon_n})$$

is not empty; clearly

$$E_1 \subseteq E_2 \subseteq \ldots \subseteq E_n \subseteq \ldots \quad \text{and} \quad E = \bigcup_{n=1}^{\infty} E_n.$$

Now since E is of second category, there exists an integer N such that E_N (and every set of higher rank) is also of second category on $|z| = 1$.

(ii) To select from E_N we now cover S with a finite net of meshes T_1, T_2, \ldots, T_m each of diameter less than $\frac{1}{4}\epsilon_N$. We treat the meshes T as closed. For a given $\mu \leqslant m$ we denote by $E_{N\mu}$ that subset of E_N for which

$$T_\mu \cap C(f, e^{i\theta}) \cap \mathscr{C}(C_{\mathscr{G}_\theta}(f, e^{i\theta})_{+\epsilon_N})$$

is not empty; since $E_N = \bigcup_{\mu \leqslant m} E_{N\mu}$, we can find a $\mu = M$ such that E_{NM} is of second category on $|z| = 1$.

(iii) We now decompose E_{NM} as follows. We choose a point a in T_M; since T_M is of diameter less than $\frac{1}{4}\epsilon_N$ and also contains a point of $\mathscr{C}(C_{\mathscr{G}_\theta}(f, e^{i\theta})_{+\epsilon_N})$ for $e^{i\theta} \in E_{NM}$, it follows that for all $e^{i\theta} \in E_{NM}$ the distance of a from $C_{\mathscr{G}_\theta}(f, e^{i\theta})$ is greater than $\frac{3}{4}\epsilon_N$. We define E_{NM1} to be the subset of E_{NM} such that the spherical distance $[f(z), a]$ between $f(z)$ and a satisfies

$$[f(z), a] > \tfrac{1}{2}\epsilon_N \qquad (4.4.1)$$

for all z in \mathscr{G}_θ with $\frac{1}{2} < |z| < 1$, E_{NM2} to be the subset of E_{NM} such that (4.4.1) is satisfied for all z in \mathscr{G}_θ with $\frac{3}{4} < |z| < 1$, and generally, E_{NMp} to be the subset of E_{NM} such that (4.4.1) is satisfied for all z in \mathscr{G}_θ with $1 - 2^{-p} < |z| < 1$. It is clear that

$$E_{NM1} \subseteq E_{NM2} \subseteq \ldots \subseteq E_{NMp} \subseteq \ldots$$

and
$$E_{NM} = \bigcup_{p=1}^{\infty} E_{NMp}. \qquad (4.4.2)$$

Since E_{NM} is of second category, there is an integer P such that E_{NMP} is of second category on $|z| = 1$; furthermore, there is an arc β on $|z| = 1$ on which E_{NMP} is dense. As $e^{i\theta}$ traverses β, the set \mathscr{G}_θ sweeps out a domain B. Let β' be an open arc contained in β. Since \mathscr{G}_θ is contained in $|z| < 1$ we can find $\delta > 0$ such that B contains the annular wedge Σ abutting on β' and bounded away from the origin by the two radial segments joining the end-points of β' to the arc of $|z| = 1 - \delta$ of the same angular length. We choose $\delta < 2^{-p}$. The set \mathscr{G}_θ, $e^{i\theta} \in E_{NMP}$, is dense in Σ and, for all $z \in \Sigma \cap \mathscr{G}_\theta$, $e^{i\theta} \in E_{NMP}$, the inequality (4.4.1) is satisfied. Hence, by continuity, $[f(z), a] > \frac{1}{2}\epsilon_N$ for all z in Σ. Since $a \in T_M$ and since the diameter of T_M is less than $\frac{1}{4}\epsilon_N$, it follows that $T_M \subseteq \mathscr{C}C(f, e^{i\theta'})$ for all $e^{i\theta'}$ in β'. But, since E_{NMP} is dense on β and therefore also on β', we can find $e^{i\theta_0} \in \beta' \cap E_{NMP}$ such that

$$T_M \cap C(f, e^{i\theta_0}) \cap \mathscr{C}(C_{\mathscr{G}_{\theta_0}}(f, e^{i\theta_0})_{+\epsilon_N})$$

is not empty, from which it follows that $T_M \cap C(f, e^{i\theta_0})$ is not empty. This contradiction proves Theorem 4.8.

As we have remarked, the most useful form of the sets \mathscr{G}_θ is that of a simple curve such as a radius or a chord terminating at $e^{i\theta}$. In Theorem 4.6 the nature of the curve γ is, to a considerable degree, unspecified, but if a specific curve is chosen with the property that $C_\gamma(f, e^{i\theta}) = C(f, e^{i\theta})$, then Theorem 4.8 asserts that the rotations† of that curve will be sets of maximum indetermination for a residual set on $|z| = 1$. A particularly interesting example is that of a Blaschke product such that every point of $|z| = 1$ is a limit point of a subsequence of the zeros of $B(z)$. Now at every point $e^{i\theta}$ of $|z| = 1$, as we shall show in Chapter 5, $C(B, e^{i\theta})$ is the closed disc $|w| \leqslant 1$. By Theorem 4.8

$$C_\rho(B, e^{i\theta}) = C(B, e^{i\theta})$$

on a residual set on $|z| = 1$. But since Fatou's theorem asserts that $C_\rho(B, e^{i\theta})$ is degenerate almost everywhere on $|z| = 1$, we see that the residual set given by Theorem 4.8 is of measure zero.

A trivial example shows that Theorem 4.8 is best possible in the sense that it may fail if the condition that $f(z)$ be continuous is relaxed to semi-continuity. Let $Z = \{z_n\}$ be a sequence of isolated points in $|z| < 1$ such that $Z' = K$, put $f(z_n) = 1$ for all n and $f(z) = 0$ elsewhere. Then $C(f, e^{i\theta}) = (0) \cup (1)$ for all θ while $C_\rho(f, e^{i\theta}) = (0)$ for all except perhaps a countable set of values of θ. The function thus defined is upper semi-continuous but does not have the maximality property on a residual set.

However, continuity, although sufficient, is not a necessary condition for the maximality property of Theorem 4.8. This theorem has recently been shown by Arsove [1] to be valid for a class of functions which includes the subharmonic functions in $|z| < 1$. This important class of functions are upper semi-continuous but not necessarily continuous.

5. Maximality theorems for an arbitrary function

It can be shown by an argument which differs only at the final stage from the proof of Theorem 4.8 that if the set \mathscr{G}_θ is a domain

† It is possible to formulate these results with families of curves which are less restrictive than the family of rotations of a fixed curve; for example, the homeomorphs of the radii under a homeomorphism of the closed disc onto itself.

generated by sweeping out along a curve λ_θ, the rotation of a curve λ_0 terminating at the point $z = 1$, with a disc of diminishing diameter tending to 0 as $z \to 1$, then the condition of continuity of f may be suppressed in Theorem 4.8. A case of special interest is that where \mathscr{G}_θ is a Stolz angle $\Delta(\theta)$. It will be sufficient to state the theorem and indicate the proof in this case.

THEOREM 4.9. *If the real or complex function $f(z)$ is defined in $|z| < 1$, but is otherwise arbitrary, and if $\Delta(\theta)$ is the Stolz angle of vertex $e^{i\theta}$ obtained by rotation of any Stolz angle $\Delta(0)$ having vertex $z = 1$, then*

$$C_{\Delta(\theta)}(f, e^{i\theta}) = C(f, e^{i\theta}),$$

except perhaps for a set of first category on $|z| = 1$.

As in the proof of Theorem 4.8 we assume the contrary, so that the set E of points of $|z| = 1$ at which

$$C_{\Delta(\theta)}(f, e^{i\theta}) \neq C(f, e^{i\theta})$$

is of second category. In the same three steps we select a subset E_0 of E, also of second category and therefore dense on an arc β of K in which a closed subset T_M of S meets $C(f, e^{i\theta})$, but is uniformly bounded away from T_M throughout an annular wedge abutting on an arc $\beta' \subset \beta$. In the previous proof we replace \mathscr{G}_θ by $\Delta(\theta)$ and observe, at stage (iii) of the selection, that every interior point of the wedge Σ is a point of some $\Delta(\theta)$, $\theta \in E_{NMP}$, so that the inequality (4.4.1) holds throughout Σ. From this it follows that $f(z)$ is uniformly bounded away from T_M in Σ so that T_M does not meet $C(f, e^{i\theta})$ for any $e^{i\theta} \subset \beta$. But this is inconsistent with E_{NMP} and therefore also E_{NM}, being dense in β since T_M meets $C(f, e^{i\theta})$ for all $e^{i\theta} \in E_{NM}$. This contradiction proves the theorem.

From this theorem we can easily deduce a theorem for arbitrary Stolz angles. Let $\{\varphi_m\}$ be a dense, countable set in the open interval $-\frac{1}{2}\pi < \varphi < \frac{1}{2}\pi$, and let

$$\eta_1 > \eta_2 > \dots > \eta_n > \dots, \quad \lim \eta_n = 0,$$

be a sequence of real numbers. We denote by $\Delta_{mn}(\theta)$ the angular domain of opening η_n bisected by the chord $\rho(\varphi_m)$ to $e^{i\theta}$ and contained in $|z| < 1$. For a given m, this sets a lower bound $n(m)$ to n,

the permitted values of η_n for φ_m being those for which $n \geqslant n(m)$. By Theorem 4.9, for given m and $n \geqslant n(m)$, the relation

$$C_{\Delta_{mn}(\theta)}(f, e^{i\theta}) = C(f, e^{i\theta})$$

is satisfied for all $e^{i\theta} \in K$ except perhaps for a set $E_{m,n}$ of first category. It follows that, for *every* pair m and $n \geqslant n(m)$, this relation is satisfied for all $e^{i\theta} \in K - E$, where

$$E = \bigcup_{\substack{1 \leqslant m < \infty \\ n(m) \leqslant n < \infty}} E_{mn}$$

is a set of first category. Now given any Stolz angle Δ having its vertex at $e^{i\theta}$, we can find m and n such that $\Delta_{mn}(\theta) \subset \Delta$, so that

$$C_\Delta(f, e^{i\theta}) \supseteq C_{\Delta_{mn}}(f, e^{i\theta}) = C(f, e^{i\theta}).$$

We have thus proved

THEOREM 4.10. *If $f(z)$ is an arbitrary one-valued (real or complex) function defined in $|z| < 1$, and if Δ is any Stolz angle in $|z| < 1$ with vertex at $e^{i\theta}$, then $C_\Delta(f, e^{i\theta}) = C(f, e^{i\theta})$ except perhaps for a set of first category on $|z| = 1$.*

This theorem, valid for an arbitrary function and a topologically 'thick' set, may be contrasted with the theorems of Fatou and Plessner (Theorems 2.4 and 8.2), valid for an analytic or meromorphic function and a metrically 'thick' set.

We must remember that in the case of an arbitrary function defined in $|z| < 1$ the circumference $K: |z| = 1$ is not a privileged curve. Analogous results to those proved above may be obtained for other curves in $|z| < 1$.† The interest of theorems on the boundary K is that they are applicable to functions subject to further restrictive conditions in $|z| < 1$ and for which K has a special status.

As we remarked at the beginning of this paragraph, Theorem 4.9 is valid, and is proved by the same argument, when the Stolz angle $\Delta(\theta)$ is replaced by a $\mathscr{G}_\theta = \Delta(\theta, \sigma)$ generated by sweeping out along the curve λ_θ with a disc whose radius σ diminishes to 0

† The theory of cluster sets of arbitrary functions, but restricted to real functions, originated with W. H. Young [1], whose first paper on the subject was published in 1908.

as fast as we please as $z \in \lambda_\theta$ tends to $e^{i\theta}$.† We can then show, by the method of § 3, that given such a domain $\Delta(\theta, \sigma)$, a curve λ_θ' may be constructed in this domain with the property that

$$C_{\lambda_\theta'}(f, e^{i\theta}) = C_{\Delta(\theta, \sigma)}(f, e^{i\theta}),$$

and, since $\sigma(t)$ may be made to decrease as rapidly as we please, the curve may be made to lie arbitrarily close to λ_θ. Replacing $\Delta(\theta)$ by $\Delta(\theta, \sigma)$, it thus follows from Theorem 4.9 that *for an arbitrary function $f(z)$ in $|z| < 1$ and for a given rotational family of curves $\{\lambda_\theta\}$ there exists for each point $e^{i\theta} \in K$, except perhaps for a set of first category on K, a curve λ_θ' arbitrarily close to λ_θ on which $C_{\lambda_\theta'}(f, e^{i\theta}) = C_{\lambda_\theta}(f, e^{i\theta})$.*

6. Symmetric maximality theorems on cluster sets of an arbitrary function

All the subsets of $C(f, e^{i\theta})$ considered up to this point have been defined with respect to a subset \mathcal{G} of $|z| \leq 1$. A particularly important subset of $C(f, e^{i\theta})$, as far as its applications in the theory of functions is concerned, is a cluster set formed, not with respect to the behaviour of $f(z)$ in the domain primarily, but with respect to its behaviour near points of $|z| = 1$ in the neighbourhood of $e^{i\theta}$. In defining the *boundary cluster set* $C_B(f, e^{i\theta_0})$ of $f(z)$ at $e^{i\theta_0}$, we form first the set

$$C(f, 0 < |\theta - \theta_0| < \eta) = \bigcup_{0 < |\theta - \theta_0| < \eta} C(f, e^{i\theta}) \qquad (4.6.1)$$

for a fixed $\eta > 0$, and then define $C_B(f, e^{i\theta_0})$ to be the intersection

$$C_B(f, e^{i\theta_0}) = \bigcap_{\eta > 0} \bar{C}(f, 0 < |\theta - \theta_0| < \eta) \qquad (4.6.2)$$

where \bar{C} is the closure of C. Now $C_B(f, e^{i\theta_0})$ is a closed subset of $C(f, e^{i\theta_0})$, but even in the analytic case it need not be connected; it is then however the union of two connected subsets

$$C_B(f, e^{i\theta_0}) = C_{Bl}(f, e^{i\theta_0}) \cup C_{Br}(f, e^{i\theta_0}), \qquad (4.6.3)$$

where the left-hand and right-hand boundary cluster sets $C_{Bl}(f, e^{i\theta_0})$ and $C_{Br}(f, e^{i\theta_0})$ are defined by (4.6.1) and (4.6.2),

† Collingwood [11].

together with the additional restrictions that $0 < \theta - \theta_0 < \eta$ and $0 < \theta_0 - \theta < \eta$, respectively.

That $C_{Br}(f, e^{i\theta})$ is a subset of $C(f, e^{i\theta})$ follows from the fact, which can be proved by the argument of Theorem 4.6 with only trifling changes of detail, that, *given any curve l tangential to $K: |z| = 1$ at $e^{i\theta}$ from the right, there is a curve γ lying in the domain between K and l and terminating at $e^{i\theta}$ such that*

$$C_\gamma(f, e^{i\theta}) = C_{Br}(f, e^{i\theta}).$$

We have now only to note that

$$C_\gamma(f, e^{i\theta}) \subseteq C(f, e^{i\theta}).$$

Similarly for $C_{Bl}(f, e^{i\theta})$.

We shall next prove† that for an arbitrary function the right (left) boundary cluster set is maximal except perhaps for a countable set of points $e^{i\theta}$.

THEOREM 4.11. *If the function $f(z)$ is defined and is one-valued (real or complex) in $|z| < 1$, then*

$$C_{Br}(f, e^{i\theta}) = C_{Bl}(f, e^{i\theta}) = C(f, e^{i\theta})$$

except perhaps for a countable set of points $e^{i\theta} \in K$.

To prove this we cover the Riemann sphere S with a succession of finite nets T_k, $k = 1, 2, ...$, each being a subdivision of its predecessor. We denote the (closed) meshes of T_k by τ_{kn}, $n = 1, 2, ..., n_k$, and we impose the condition that all of them are of diameter less than $1/k$. By any convenient ordering, we arrange all the meshes τ_{kn} of this sequence $\{T_k\}$ of lattices into a sequence $\{\tau_j\}$. For each index j we consider the set E_j of those points $e^{i\theta} \in K$ such that the mesh τ_j contains a point of $C(f, e^{i\theta})$ but does not meet the set $C_{Br}(f, e^{i\theta})$. Denote by E_R the set of points $e^{i\theta}$ for which $C(f, e^{i\theta}) \cap \mathscr{C}C_{Br}(f, e^{i\theta})$ is not empty; clearly $E_R = \bigcup_j E_j$. If $e^{i\theta} \in E_j$, then $e^{i\theta}$ is the left-hand end-point of some arc $A(\theta) \subset K$ such that $C(f, e^{i\varphi})$ does not meet the mesh τ_j for any point $e^{i\varphi} \in A(\theta)$, for otherwise, τ_j would meet the set $C_{Br}(f, e^{i\theta})$, contrary to the hypothesis that $e^{i\theta} \in E_j$. Since the arc $A(\theta)$ cannot contain points of E_j, the set E_j is at most countable, so that E_R is at most countable. Likewise, the corresponding set E_L in which

† Collingwood [11].

$C(f, e^{i\theta}) \cap \mathscr{C}C_{Bl}(f, e^{i\theta})$ is not empty is at most countable, and the theorem is proved.†

The following definitions will also be useful. The *right-hand cluster set* $C_R(f, e^{i\theta})$ is the set of points α such that

$$\alpha = \lim_{n \to \infty} f(r_n e^{i\theta_n}),$$

where $\lim r_n = 1$ and $\lim \theta_n = \theta$, with $\theta_n \leqslant \theta_{n+1} \leqslant \dots$. The *left-hand cluster set* $C_L(f, e^{i\theta})$ is defined in the same way except that $\theta_n \geqslant \theta_{n+1} \geqslant \dots$.

We noted at the beginning of this paragraph that we can find a path γ terminating at $e^{i\theta}$, lying wholly to the right and arbitrarily near an arc of K, having $e^{i\theta}$ as its end-point, such that $C_\gamma(f, e^{i\theta}) = C_{Br}(f, e^{i\theta})$. Therefore

$$C_{Br}(f, e^{i\theta}) \subset C_R(f, e^{i\theta}), \quad \text{for} \quad C_\gamma(f, e^{i\theta}) \subset C_R(f, e^{i\theta});$$

similarly, $C_{Bl}(f, e^{i\theta}) \subset C_L(f, e^{i\theta})$, and we have the

COROLLARY. *If $f(z)$ is defined and is one-valued (real or complex) in $|z| < 1$, then* $C_R(f, e^{i\theta}) = C_L(f, e^{i\theta}) = C(f, e^{i\theta})$

except perhaps for a countable set of points $e^{i\theta} \in K$.

7. Bagemihl's theorem on ambiguous points of an arbitrary function

We continue our development of what we have called the intrinsic properties of cluster sets by examining the relationship between the curvilinear cluster sets defined by two arbitrary arcs γ_1 and γ_2 ending at a point $e^{i\theta}$ on $K: |z| = 1$. The principal result of this section‡ that the set of *ambiguous* points $e^{i\theta}$ on K where there exist two arcs γ_1 and γ_2 at $e^{i\theta}$ such that the intersection

$$C_{\gamma_1}(f, e^{i\theta}) \cap C_{\gamma_2}(f, e^{i\theta}) = \varnothing$$

is at most countable is again valid for an arbitrary function $f(z)$ defined in $|z| < 1$. Nevertheless, this result (Theorem 4.12), like the theorems for arbitrary functions proved earlier in this

† Further symmetry theorems have recently been obtained by Ulysses Hunter [1, 2].

‡ Bagemihl [1].

chapter, has wide application in the case that $f(z)$ is meromorphic in $|z| < 1$. We require first the

LEMMA. *For every set M in D: $|z| < 1$, there is a set K_0 on K: $|z| = 1$, whose complement $\mathscr{C}K_0$ on K is at most countable, such that for every $e^{i\theta} \in K_0$ and for every pair of arcs γ_1 and γ_2 terminating at $e^{i\theta}$, either γ_1 and γ_2 both intersect M or both intersect $D - M$.*

For every pair of rational numbers α and ρ such that

$$0 < \alpha < \tfrac{1}{2}\pi, \quad 0 < \rho < 1, \tag{4.7.1}$$

we say that $e^{i\theta}$ belongs to the set $E_{\alpha,\rho}$ if there exist arcs γ_1 and γ_2 at $e^{i\theta}$ with their end-points ζ and ζ', respectively, lying in M such that

$$|\zeta| = |\zeta'| = \rho, \quad \theta - \tfrac{1}{4}\pi < \arg\zeta' < \theta + \tfrac{1}{4}\pi - \alpha < \arg\zeta < \theta + \tfrac{1}{4}\pi, \tag{4.7.2}$$

$$\rho < |z| < 1, \quad \theta - \tfrac{1}{4}\pi < \arg z < \theta + \tfrac{1}{4}\pi \tag{4.7.3}$$

for every z in $\gamma \cup \gamma'$ which is distinct from ζ or ζ', and

$$\gamma \subset M, \quad \gamma' \subset D - M. \tag{4.7.4}$$

We define the set $E'_{\alpha,\rho}$ in the same way except that we replace condition (4.7.4) by

$$\gamma \subset D - M, \quad \gamma' \subset M. \tag{4.7.5}$$

The sets $E_{\alpha,\rho}$ and $E'_{\alpha,\rho}$ are isolated and therefore are at most countable. Therefore the union $\bigcup_{\alpha,\rho} (E_{\alpha,\rho} \cup E'_{\alpha,\rho})$ taken over all rational numbers α and ρ satisfying (4.7.1) is also at most countable, so that if K_0 is the set of all $e^{i\theta}$ not belonging to any $E_{\alpha,\rho}$ or $E'_{\alpha,\rho}$, then $\mathscr{C}K_0$ is at most countable on K.

Let $e^{i\theta}$ be an arbitrary point of K_0, and let γ_1 and γ_2 be two arcs terminating at $e^{i\theta}$. We may assume that $\gamma_1 \cap \gamma_2 = \varnothing$, for otherwise the conclusion of the lemma follows at once. If either of the radii $\arg z = \theta \pm \tfrac{1}{4}\pi$ intersects γ_1, we denote by γ'_1 that subarc running from $e^{i\theta}$ to the first point of intersection, while if neither radius intersects γ_1, we set $\gamma_1 = \gamma'_1$; the arc γ'_2 is derived from γ_2 in a similar way. We can find a rational number ρ, $0 < \rho < 1$, such that the circle $|z| = \rho$ intersects both γ'_1 and γ'_2 and, at the same time, contains in its interior the end-points of γ'_1 and γ'_2. Denote by ζ the first point of intersection of γ'_1 with $|z| = \rho$ starting from

$e^{i\theta}$, and let γ denote the subarc of γ'_1 so formed; similarly, we define a point ζ' on $|z| = \rho$ and γ'_2 together with an arc γ' running from $e^{i\theta}$ to ζ'. Now $\zeta \neq \zeta'$, and we may assume that $\arg \zeta' < \arg \zeta$. Now on the shorter arc of $|z| = \rho$ determined by the points $\rho e^{i(\theta \pm \frac{1}{4}\pi)}$ there exists a point β such that $\alpha = \theta + \frac{1}{4}\pi - \arg\beta$ is a rational number between 0 and $\frac{1}{2}\pi$. Thus, for every z in $\gamma \cup \gamma'$ and distinct from ζ or ζ' we have (4.7.1), (4.7.2) and (4.7.3). Since $e^{i\theta} \in K_0$, neither (4.7.4) nor (4.7.5) can hold, so that either γ and γ' both intersect M or both intersect $D - M$, which implies that the same is true of γ_1 and γ_2. This proves the lemma.

THEOREM 4.12. *If $f(z)$ is a complex-valued function defined in $|z| < 1$, then the set of ambiguous points $e^{i\theta}$ on $|z| = 1$ with the property that there exist two arcs γ_1 and γ_2 such that*

$$C_{\gamma_1}(f, e^{i\theta}) \cap C_{\gamma_2}(f, e^{i\theta}) = \varnothing$$

is at most countable.

Let $\{H_m\}$ be a basis for the sphere S consisting of all open discs having rational† centres and radii; the open sets on S which can be expressed as a finite union of the circles $\{H_m\}$ form a countable set which we denote by $\{G_n\}$. For each n we define the set

$$S_n = \{z : |z| < 1; f(z) \in \mathscr{C}G_n\}. \tag{4.7.6}$$

By our lemma there exists a set $K_0^{(n)}$ on $|z| = 1$ with $\mathscr{C}K_0^{(n)}$ at most countable such that, if $e^{i\theta} \in K_0^{(n)}$ and if γ'_1 and γ'_2 are any two arcs terminating at $e^{i\theta}$, then either γ'_1 and γ'_2 both intersect S_n or both intersect $D - S_n$. We set $K_0 = \bigcap_n K_0^{(n)}$, and observe that $\mathscr{C}K_0$ is at most countable.

Let $e^{i\theta} \in K_0$ and let γ_1 and γ_2 be any two arcs at $e^{i\theta}$. If

$$C_{\gamma_1}(f, e^{i\theta}) \cap C_{\gamma_2}(f, e^{i\theta}) = \varnothing, \tag{4.7.7}$$

there is an integer n such that

$$C_{\gamma_1}(f, e^{i\theta}) \subset G_n \tag{4.7.8}$$

and
$$C_{\gamma_2}(f, e^{i\theta}) \subset S - \bar{G}_n. \tag{4.7.9}$$

Now let $\gamma_{11} \supset \gamma_{12} \supset \dots \supset \gamma_{1,k} \supset \dots$ be a sequence of subarcs of γ_1, each of which terminates at $e^{i\theta}$, and whose end-points converge to

† In terms of some projection of S onto the plane.

$e^{i\theta}$; let $\{\gamma_{2,k}\}$ denote a similar sequence of subarcs of γ_2. For every integer k, $\gamma_{1,k}$ and $\gamma_{2,k}$ must both intersect S_n or else must both intersect $D - S_n$, since $e^{i\theta} \in K_0^{(n)}$. Thus there must exist an increasing sequence $\{k_j\}$ of integers such that either γ_{1,k_j} and γ_{2,k_j} both intersect S_n for every j, or else, for every j, both intersect $D - S_n$. By the definition of S_n, the first alternative implies that $C_{\gamma_1}(f, e^{i\theta}) \cap (S - G_n)$ is not empty, which contradicts (4.7.8), while the second alternative implies that $C_{\gamma_2}(f, e^{i\theta}) \cap \bar{G}_n$ is not empty, which contradicts (4.7.9). Consequently (4.7.7) cannot hold for any $e^{i\theta} \in K_0$; since the complement of K_0 on $|z| = 1$ is at most countable, Theorem 4.12 is proved.

We have already seen that the property (4.7.7) can hold for the bounded analytic function $w = f(z) = e^{(z+1)/(z-1)}$ with $e^{i\theta} = 1$, for if γ_1 is the radius $\arg z = 0$ and γ_2 the upper semicircle of $|z - \tfrac{1}{2}| = \tfrac{1}{2}$, then $C_{\gamma_1}(f, 1)$ is the point $\{0\}$, while $C_{\gamma_2}(f, 1)$ is a set lying on the circle $|w| = e^{-1}$. Fairly simple examples have been constructed† to show that, given a countable set E on $|z| = 1$, there exists an analytic function having the property (4.7.7) at each point of E, so that Theorem 4.12 cannot be improved, even by imposing the condition of regularity on $f(z)$. On the other hand, in Chapter 5, § 1, we shall give sufficient conditions for the set of ambiguous points to be empty.

Theorem 4.12 is also best possible in another sense. For, the set of points of K at which the cluster sets of f on any three curves have an empty intersection is not necessarily countable. An example, for which the curves are chords, was given by Jarník [1]. Other examples, including one for which the intersection of any three chordal cluster sets is empty at every point of K, have been given by Bagemihl, Piranian and Young [1] who also point to a number of open questions in this area. The theorem is also essentially two-dimensional in the sense that it has no useful analogue for a function defined in a sphere.

8. Normal functions. More general classes of functions

The so-called normal functions form a restricted but important class of meromorphic functions defined in the unit disc. This class

† Bagemihl and Seidel [4]; Lohwater and Piranian [2].

includes several classes with which we are already familiar, such as the class of bounded functions, the class of functions omitting three values and the class of functions mapping the disc onto a surface of bounded spherical area. But since the relation is one of inclusion, the conditions under which properties are shared need to be defined. The subject, which originated with Noshiro [1], was extensively developed by Lehto and Virtanen [1], and the theme was picked up later by others (e.g. Bagemihl [3], Bagemihl and Seidel [6, 7], Gavrilov [3, 4], and others). Lack of space forces us to limit ourselves here, in spite of the interest in normal functions, to the definition of this class and by way of illustration, to mention a theorem of Seidel [2] which is of particular interest from the point of view of cluster-set theory.

By definition, a function $f(z)$, meromorphic in $|z| < 1$, is *normal* if the family $\{f(T(z))\}$ is normal in the sense of Montel, where $T(z)$ is an arbitrary conformal mapping of $|z| < 1$ onto itself. Clearly, the definition may be extended by replacing the disc by an arbitrary simply connected domain D or the universal covering surface of a more general domain. As a basic criterion of normality we have the following (Lehto and Virtanen [1], Noshiro [4]):

A non-constant function $f(z)$, meromorphic in $|z| < 1$, is normal in $|z| < 1$ if and only if there is a constant C such that

$$\frac{|f'(z)|}{1 + |f(z)|^2} < \frac{C}{1 - |z|^2}.$$

The theorem of Seidel is the following, which we state without proof since his proof, which was given originally for a bounded function [2], is immediately applicable to a normal function.

THEOREM 4.13. *If $f(z)$ is meromorphic and normal in $|z| < 1$, and if λ_1 and λ_2 are any two non-intersecting Stolz paths in $|z| < 1$ terminating at $e^{i\theta}$, then either $C_{\lambda_1}(f, e^{i\theta}) = C_{\lambda_2}(f, e^{i\theta})$, or else $C(f, e^{i\theta})$ contains a disc.*

In addition, it must be pointed out that Doob [7] has been able to unify the study of other important classes of analytic functions in the theory by replacing the notion of angular cluster set by that of 'fine' boundary function or 'fine' limit in the Cartan–Brelot–Naïm fine topology. The new theorems are sometimes not strict generalizations of the classical versions, because these classical

versions do not always possess a conformally invariant form, and one may expect much development in this area.

There also exist extensive theories of quasi-conformal, and, more generally, pseudo-analytic, functions (for earlier references, see Noshiro [4], as well as Szilárd [1–3] who was one of the earlier workers in the field of non-analytic complex function theory).

9. Generalizations

As might be expected because of their topological character, many of the results of this chapter can be given a more general formulation, either as regards the dimension of the space for which they are formulated or its topology. In particular, the maximality theorems of §§4 and 6 have been generalised by Weston [1]. To discuss these developments, interesting as they are, would take us too far from our main theme in this monograph, which is the theory of cluster sets as it is applicable to the theory of functions. Because of our immediate interest in the application of the theory to various aspects of the boundary behaviour of analytic functions, and because of limitations of space, we have confined our discussions to the two-dimensional case. But we may point out that the maximality Theorem 4.8 is valid for a continuous function in an n-ball, in particular, the euclidean sphere, the proof requiring only minor changes from that of Theorem 4.8. The theorem is thus applicable to quasi-conformal mappings in space (Gehring [3], Väisälä [1], Church [3]).

CHAPTER 5

CLUSTER SETS OF FUNCTIONS ANALYTIC IN THE UNIT DISC

1. The Iversen–Beurling theorem

It is the relationship between two or more of the cluster sets at a point which determines, to a large extent, the usefulness of the theory of cluster sets in the study of the boundary behaviour of meromorphic functions. As we have already pointed out, certain classical developments in the theory of functions have indicated the form which part of the theory of cluster sets must assume; we mention again the theorems of Weierstrass, Picard, and Iversen in the case of the isolated boundary point and the theorems of Fatou and Lindelöf, and others, in the case of the continuous boundary.

We begin by proving a relationship between the frontier $\mathrm{Fr}\, C(f, e^{i\theta})$ of $C(f, e^{i\theta})$ and $C_B(f, e^{i\theta})$ which depends for its validity, as an example shows, upon the regularity (or near-regularity) of $f(z)$ and so falls outside the general scope of the theorems described in Chapter 4. We denote by R_1 and R_2 the rays $\arg z = \frac{1}{6}\pi$ and $\arg z = -\frac{1}{6}\pi$ respectively, and let $f(z)$ be bounded in the domain $G: 0 < |z| < R_0$, $|\arg z| < \frac{1}{6}\pi$. We denote by M the quantity

$$\overline{\lim_{z' \to 0}} (\overline{\lim_{z \to z'}} |f(z)|), \tag{5.1.1}$$

where $z' \neq 0$ is a point of $R_1 \cup R_2$, and we consider the function $F(z) = f(z)/M$, where, for reasons which will be obvious when we state the theorem to be proved, we may assume that $M \neq 0$.

We assert that $\overline{\lim_{z \to 0}} |F(z)| \leqslant 1$.

Let us assume to the contrary that $\overline{\lim_{z \to 0}} |F(z)| = m > 1$ and force a contradiction. Given $\epsilon_1 > 0$, it follows that we may find a number ρ_{ϵ_1} such that in the common part G_1 of G and $|z| < 2\rho_{\epsilon_1}$ we have $\overline{\lim_{z \to z'}} |F(z)| < 1 + \epsilon_1$ for z' on that part of $R_1 \cup R_2$ inside $|z| < 2\rho_{\epsilon_1}$. For any point z_0 in the common part G_1' of G and

$|z| < \rho_{\epsilon_1}$, we consider the equilateral triangle Δ_1 with z_0 as centroid and at least one of its edges on $R_1 \cup R_2$. If we set $\zeta = z - z_0$, we have $F(z) = F_1(\zeta)$, say, and we see that the function

$$g(\zeta) = F_1(\zeta)\,F_1(\omega\zeta)\,F_1(\omega^2\zeta),$$

where ω is a complex cube root of unity, is analytic and bounded

$$|g(\zeta)| < (1+\epsilon_1)\,M_0^2$$

on the boundary of Δ_1; here M_0 denotes any bound on $|F(\zeta)|$ in G. Hence $|g(\zeta)| < M_0^2$ inside Δ_1, so that, by setting $\zeta = 0$, we have $|F(z_0)| < M_0^{\frac{2}{3}}$. Since z_0 was an arbitrary point of R_{ϵ_1}, we have that $|F(z)| < M_0^{\frac{2}{3}}$ in G_1'.

Now for $\epsilon_2 < \epsilon_1$ we may find a number $\rho_{\epsilon_2} < \frac{1}{2}\rho_{\epsilon_1}$ such that, in the common part G_2 of Δ_1 and $|z| < 2\rho_{\epsilon_2}$, $\varlimsup\limits_{z \to z'} |F(z)| < 1 + \epsilon_2$ on that part of $R_1 \cup R_2$ bounding G_2. Repeating our argument for an arbitrary z_0 in that part G_2' of G common to $|z| < \rho_{\epsilon_2}$, we have $|F(z)| < M_0^{(\frac{2}{3})^2}$ for all z in G_2'. Now if $m > 1$, we can, after a finite number of such repetitions, find an integer n and a number ρ_{ϵ_n} such that $|F(z)| < M_0^{(\frac{2}{3})^n}$ in the region $|\arg z| < \frac{1}{6}\pi$, $0 < |z| < \rho_{\epsilon_n}$, where n has the property that $1 < M_0^{(\frac{2}{3})^n} < m$. This contradicts our assumption and proves our assertion that $\varlimsup\limits_{z \to 0} |F(z)| \leqslant 1$, or, stated in terms of $f(z)$, that

$$\varlimsup_{z \to 0} |f(z)| \leqslant \varlimsup_{z' \to 0} (\varlimsup_{z \to z'} |f(z)|).$$

If we map the domain $|\arg z| < \frac{1}{6}\pi$, $0 < |z| < R_0$ conformally onto the unit circle, we may state the result as

THEOREM 5.1. *If $f(z)$ is analytic and bounded in $|z| < 1$, and if $e^{i\theta_0}$ is an arbitrary point of $|z| = 1$, then*

$$\varlimsup_{z \to e^{i\theta_0}} |f(z)| \leqslant \varlimsup_{\theta \to \theta_0} (\varlimsup_{z \to e^{i\theta}} |f(z)|). \tag{5.1.2}$$

The right-hand side of (5.1.2) suggests a relationship between that expression and the boundary cluster set $C_B(f, e^{i\theta_0})$ of $f(z)$ at $e^{i\theta_0}$, and we may establish this relationship in the following form†

† Iversen [2, 3]; see also Seidel [1] and Doob [1].

THEOREM 5.2. *If $f(z)$ is meromorphic in $|z| < 1$, then, at every point $e^{i\theta}$ of $|z| = 1$,* $\quad \mathrm{FR}\, C(f, e^{i\theta}) \subseteq C_B(f, e^{i\theta}),$ \qquad (5.1.3)

where $\mathrm{FR}\, C(f, e^{i\theta})$ *denotes the frontier of* $C(f, e^{i\theta})$.

We suppose that there is a point $e^{i\theta_0}$ at which (5.1.3) fails, i.e. for some $e^{i\theta_0}$ there is a point $w_0 \in \mathrm{FR}\, C(f, e^{i\theta_0})$ not in $C_B(f, e^{i\theta_0})$. Then there is a $\delta > 0$ such that w_0 is at a distance greater than δ from $C_B(f, e^{i\theta_0})$. Now w_0 is a frontier point of $C(f, e^{i\theta_0})$, so that there exists a point $w_1 \in \mathscr{C}C(f, e^{i\theta_0})$ with $|w_1 - w_0| < \frac{1}{2}\delta$. We remark that the distance between w_1 and any point of $C_B(f, e^{i\theta_0})$ must be greater than $\frac{1}{2}\delta$. Now the function $F(z) = [f(z) - w_1]^{-1}$ must be bounded in some neighbourhood of $e^{i\theta_0}$, so that Theorem 5.1 applied to $F(z)$ implies that the maximum distance from any finite point to $C(F, e^{i\theta_0})$ is not greater than the maximum distance from that point to $C_B(f, e^{i\theta_0})$, so that

$$\frac{1}{|w_0 - w_1|} \leqslant \frac{2}{\delta},$$

which contradicts the fact that we can choose w_1 so that $|w_1 - w_0| < \frac{1}{2}\delta$. Thus we have proved Theorem 5.2.

A conformal mapping onto a Jordan domain \mathscr{D} gives, as an immediate corollary of Theorem 5.2, an important theorem due to Iversen, namely

THEOREM 5.2.1. *If $f(z)$ is meromorphic in the domain \mathscr{D} bounded by a Jordan curve γ then, for any point $z_0 \in \gamma$,*

$$\mathrm{FR}\, C_{\mathscr{D}}(f, z_0) \subset C_{\gamma}(f, z_0).$$ \qquad (5.1.4)

The topological relation (5.1.4) has applications in the theory of functions.

Let γ_1 and γ_2 be open arcs of γ to the right and left of z_0, their common end-point, so that

$$C_{\gamma}(f, z_0) = C_{\gamma_1}(f, z_0) \cup C_{\gamma_2}(f, z_0).$$

Since $C_{\gamma}(f, z_0)$ is closed, its complement

$$\mathscr{C}C_{\gamma}(f, z_0) = \mathscr{C}C_{\gamma_1}(f, z_0) \cap \mathscr{C}C_{\gamma_2}(f, z_0)$$

is either empty or consists of a finite or countable set of open components. Since $C_{\mathscr{D}}(f, z_0)$ is a continuum it follows from (5.1.4) that,

for any such component δ, either $\delta \subset C_{\mathscr{D}}(f, z_0)$ or $\delta \cap C_{\mathscr{D}}(f, z_0) = \varnothing$. In particular, if $C_{\gamma_1}(f, z_0) \cap C_{\gamma_2}(f, z_0) = \varnothing$ *the unique component δ whose frontier contains points of both $C_{\gamma_1}(f, z_0)$ and $C_{\gamma_2}(f, z_0)$ is contained in $C_{\mathscr{D}}(f, z_0)$.*† If we now suppose that $z_0 \in K$ and γ_1, γ_2 are in $|z| < 1$, this leads at once to a generalization of Theorem 2.3.1 which it contains as a special case. For, if $C_{\gamma_1}, C_{\gamma_2}$ are both degenerate and distinct, $C_{\mathscr{D}}(f, z_0)$ must be total. More generally, if $C(f, z_0)$ is non-dense, and *a fortiori* if $C(f)$ is non-dense, as in the case of a univalent function, for example, $C_{\gamma_1}(f, z_0) \cap C_{\gamma_2}(f, z_0) \neq \varnothing$, or, since γ_1, γ_2 are arbitrary Jordan curves, z_0 is not an ambiguous point for $f(z)$ in the sense of Theorem 4.12.

The validity of the relation (5.1.3) for a function defined in $|z| < 1$ depends rather heavily on a condition like regularity or near-regularity, as we might surmise from the proof of Theorem 5.1, where the proof depended on the regularity of the product $F_1(\zeta) F_1(\omega\zeta) F_1(\omega^2\zeta)$. We show by means of an example that (5.1.3) need not hold in the case of a light interior mapping. The function

$$w = g(z) = x + iy(1-x)$$

maps the interior of the unit square $ABCD$, where $A = 0$, $B = i$, $C = 1+i$, $D = 1$, in a topological way onto the interior of the triangle PQR, where $P = 0$, $Q = i$, $R = 1$, in the w-plane. The interesting property of $g(z)$, in addition to being light interior, is that it collapses the side CD of the square into a single point R of the triangle. Thus, by means of the inverse function

$$z = g^{-1}(w) = u + i\frac{v}{1-u},$$

which is light interior in the triangle PQR, it is possible to make a single frontier point 'open up' into a side CD of the image domain $ABCD$. Since $ABCD$ and PQR may each be mapped conformally onto a unit circle, and since the composition of an analytic function and a light interior mapping is light interior, we have shown the existence of a one-to-one light interior mapping $w = h(z)$ of $|z| < 1$ onto $|w| < 1$ with the property that $C(h, 1)$ is a semicircle $|w| = 1$, $\operatorname{Im} w \geqslant 0$, say. It is clear that no analytic

† This result was first proved by Hössjer [1].

FUNCTIONS ANALYTIC IN THE UNIT DISC 93

or quasi-conformal mapping of $|z| < 1$ can effect such a mapping, although it is of interest to note that $w = h(z)$ is an *harmonic mapping*, i.e. both real and imaginary parts of $h(z)$ are harmonic but are not necessarily conjugates. For the example just constructed, $C_B(h, 1) = \{1\} \cup \{-1\}$, so that (5.1.3) does not hold. Actually, the failure of the relation (5.1.3) in this and similar cases is not due to the fact that the product of two functions fails to belong to that class, as is generally the case with the class of light interior mappings in $|z| < 1$, for (5.1.3) is found to be valid for the case of quasi-conformal functions, but rather, from the failure of the existence of a strong form of the maximum principle for the class of functions. For functions analytic in $|z| < 1$ we state this principle as

THEOREM 5.3. *If $f(z)$ is analytic and bounded, $|f(z)| < M$, in $|z| < 1$, and if the radial limit values $f(e^{i\theta})$ are in modulus not greater than $m < M$ almost everywhere on $|z| = 1$, then $|f(z)| < m$ everywhere in $|z| < 1$, unless $f(z)$ is a constant of modulus m.*

Since $f(z)$ can be represented by a Poisson integral in terms of the values $f(e^{i\theta})$ almost everywhere on $|z| = 1$, we have

$$f(z) = \frac{1}{2\pi} \int_{E_t} f(e^{it}) \frac{1-r^2}{1+r^2-2r\cos(\theta-t)} dt,$$

where E_t is a set of measure 2π on $|z| = 1$.

Thus, if E_t is the set in which $|f(e^{i\theta})| \leqslant m$,

$$|f(z)| \leqslant \frac{1}{2\pi} \int_{E_t} |f(e^{it})| \frac{1-r^2}{1+r^2-2r\cos(\theta-t)} dt$$

$$\leqslant \frac{m}{2\pi} \int_0^{2\pi} \frac{1-r^2}{1+r^2-2r\cos(\theta-t)} dt = m,$$

from which it follows that $|f(z)| < m$ in $|z| < 1$, unless $f(z)$ is a constant of modulus m.

By means of the function $w = h(z)$ constructed above, together with a conformal mapping of $|w| < 1$ onto an appropriate domain D_t in the t-plane, it is easy to construct a light interior mapping $t = H(z)$ for which the stronger maximum principle fails to hold, even when the exceptional set permitted by Theorem 5.3 consists of the single point $z = 1$.

2. Extensions of the Schwarz reflexion principle

The fact that an extended maximum principle holds with no regard to the behaviour on a set of measure zero in Theorem 5.3 suggests that we seek stronger forms of Theorems 5.1 and 5.2. This we do by first extending the classical Schwarz reflexion principle:

If $f(z)$ is analytic and bounded, $|f(z)| < 1$, in $|z| < 1$, and if $f(e^{i\theta})$ is continuous with $|f(e^{i\theta})| = 1$ on an arc $A(\theta_1 < \theta < \theta_2)$ of $|z| = 1$, then $f(z)$ may be continued analytically across the arc A by means of the reflexion principle

$$f\left(\frac{1}{z}\right) = \frac{1}{\bar{f}(\bar{z})}. \qquad (5.2.1)$$

In most of the important generalizations of the reflexion principle, the condition that $f(z)$ be continuous on A is relaxed by assuming first that the radial limits of $f(z)$ exist with modulus 1 *almost everywhere* on A. As simple examples show, e.g.

$$f(z) = e^{(z+1)/(z-1)}, \quad A: -\theta_0 < \theta < \theta_0,$$

this new condition is not in itself sufficient that $f(z)$ be analytic on A, and a further condition must be imposed. Thus, if $f(z)$ has at least one singularity on A, the nature of the violated condition is equivalent to a description, in some sense, of the behaviour of $f(z)$ in the vicinity of the singularity on A.

As a first extension of the reflexion principle, we show that if $f(e^{i\theta})$ exists with modulus 1 almost everywhere on A and if P is a singularity of $f(z)$ on A, then $C(f, P)$ contains the value 0. For otherwise there would exist a number δ, $0 < \delta < 1$, such that $\delta < |f(z)| < 1$ for all z in that part of some circle $V_{\rho_1}: |z - P| < \rho_1 < 1$ lying in $|z| < 1$. Since the radial limits of $f(z)$ exist with modulus 1 almost everywhere on A, it is possible to find $\rho < \rho_1$ such that the limits of $f(z)$ exist with modulus 1 as $|z| \to 1$ along both arcs of the circle $C_\rho: |z - P| = \rho$. The function

$$F(z) = \frac{1}{2\pi i} \int_{C_\rho} \frac{\varphi(\zeta)}{\zeta - z} d\zeta, \qquad (5.2.2)$$

where $\varphi(z)$ is defined to be $f(z)$ for $|z| < 1$ and $1/[\bar{f}(1/\bar{z})]$ for $|z| > 1$, is analytic in $|z - P| < \rho$. The circles $|z| = 1 \pm \epsilon_1$ decompose

$|z - P| < \rho$ into three regions, and by taking ϵ_1 small enough, it is easy to show that $F(z)$ coincides with $f(z)$ for z in the common part of V_ρ and $|z| < 1$, and with $1/[\bar{f}(1/\bar{z})]$ for z in the common part of V_ρ and $|z| > 1$. Hence $F(z)$ provides an analytic continuation of $f(z)$ across some arc of A containing P; in particular, $f(z)$ is analytic at P, contrary to hypothesis, so that $C(f, P)$ must contain the point $w = 0$. By a linear transformation, we have

THEOREM 5.4.† *Let* $w = f(z)$ *be analytic and bounded,* $|f(z)| < 1$, *in* $|z| < 1$, *and let* $f(e^{i\theta})$ *have modulus* 1 *almost everywhere on some arc* $A(\theta_1 < \theta < \theta_2)$ *of* $|z| = 1$. *Then, at every singularity* P *of* $f(z)$ *on* A, *the cluster set* $C(f, P)$ *is the closed disc* $|w| \leq 1$.

For if α is an arbitrary complex number with $|\alpha| < 1$, then the function

$$g(z) = \frac{f(z) - \alpha}{1 - \bar{\alpha}f(z)}$$

has radial limits of modulus 1 almost everywhere on the arc A and has a singularity at P whenever $f(z)$ does. Therefore, the value 0 is in $C(g, P)$, whence $\alpha \in C(f, P)$. If we recall that $C(f, P)$ is closed and connected, we see that any value α with $|\alpha| = 1$ must also belong to $C(f, P)$, and the theorem is proved.

Since a Blaschke product $B(z)$ has radial limits of modulus 1 almost everywhere, and since any limit point of zeros of $B(z)$ must be a singularity, we have

COROLLARY 1. *If* $e^{i\theta_0}$ *is a limit point of zeros of a Blaschke product, then* $C(B, e^{i\theta_0})$ *is the closed disc* $|w| \leq 1$.

We shall make use of an extended form of Theorem 5.4, but before stating the extension, it will be convenient to make some preliminary remarks, including a geometric interpretation of the Cauchy integral. If $f(z)$ is analytic in $|z| \leq 1$, the point

$$f(0) = \frac{1}{2\pi} \int_0^{2\pi} f(e^{i\theta}) \, d\theta \qquad (5.2.3)$$

cannot lie outside the convex hull‡ of the curve given by $w = f(e^{i\theta})$, for the point $f(0)$ in (5.2.3) represents the centre of gravity of a certain mass distribution on the curve. We shall

† Seidel [1].
‡ The *convex hull* of a set M is the intersection of all convex sets containing M.

show now that this interpretation may be carried over to the case that $f(z)$ need not be analytic on $|z| = 1$; we assume, however, that $f(z)$ is analytic with $|f(z)| < M$ in $|z| < 1$. Since the radial limits of $f(z)$ exist almost everywhere, it is possible to find an open set G on K, of measure less than any given $\epsilon > 0$, such that for $e^{i\theta}$ in $K - G = \mathscr{C}G$, the limit

$$\lim_{r \to 1} f(r\,e^{i\theta}) \tag{5.2.4}$$

exists uniformly. Let G_r be the radial projection of the set G on $|z| = r$, and let $\mathscr{C}G_r$ denote the complement of G_r on $|z| = r$. We form next a continuous function $g(r\,e^{i\theta})$ which is a linear function of θ on each interval of the open set G_r, and which coincides with $f(r\,e^{i\theta})$ on $\mathscr{C}G_r$. Thus we have

$$f(0) = \frac{1}{2\pi} \int_0^{2\pi} f(r\,e^{i\theta})\,d\theta = \frac{1}{2\pi} \int_0^{2\pi} g(r\,e^{i\theta})\,d\theta$$

$$+ \frac{1}{2\pi} \int_0^{2\pi} [f(r\,e^{i\theta}) - g(r\,e^{i\theta})]\,d\theta,$$

and we write $f(0) = w_1 + w_2$, where

$$w_1 = \lim_{r \to 1} \frac{1}{2\pi} \int_0^{2\pi} g(r\,e^{i\theta})\,d\theta;$$

$$w_2 = \lim_{r \to 1} \frac{1}{2\pi} \int_0^{2\pi} [f(r\,e^{i\theta}) - g(r\,e^{i\theta})]\,d\theta.$$

Now it is clear that w_1 lies in the convex hull of the set of points (5.2.4) for $e^{i\theta} \in \mathscr{C}G$, and, since $|f(r\,e^{i\theta}) - g(r\,e^{i\theta})| < 2M$, we have that $|w_2| < 2\epsilon M$.

After these preliminaries, we now prove the following theorem of Carathéodory [4].

THEOREM 5.5. *Let $f(z)$ be analytic and bounded in $|z| < 1$, and, for every $e^{i\theta}$ of an arc $A(\theta_1 < \theta < \theta_2)$ except for a set E_z of measure zero let the radial limit $f(e^{i\theta})$ lie in a set W in the w-plane. Then, for every point P of A, the cluster set $C(f, P)$ is contained in the closure of the convex hull of W.*

For any point z_0 in $|z| < 1$, the function $z = (z_0 - \zeta)/(1 - \bar{z}_0\zeta)$ maps $|z| \leqslant 1$ conformally onto $|\zeta| \leqslant 1$; the arc A is transformed into an arc A_ζ of $|\zeta| = 1$ whose length we may write as $2\pi - \frac{1}{2}\epsilon$.

The set E_z is mapped onto a set E_ζ on $|\zeta| = 1$ which is also of measure zero. We consider the function

$$F(\zeta) = f\left(\frac{z_0 - \zeta}{1 - \bar{z}_0 \zeta}\right),$$

which is analytic and bounded in $|\zeta| < 1$. We can cover the set E_ζ and the complement of A_ζ with an open set G in such a way that the sum of the lengths of the open intervals forming G is less than ϵ. Applying to $F(\zeta)$ the remarks preceding the statement of Theorem 5.5, we have that

$$f(z_0) = F(0) = w_1 + w_2,$$

where w_1 lies in the convex hull of the set W and where w_2 satisfies $|w_2| < 2\epsilon M$, where M is a bound on $|f(z)|$. To complete the proof of Theorem 5.5 we need only observe that ϵ tends to zero with $|P - z_0|$.

As a first application of Theorem 5.5 we prove the following extension of Theorem 5.1.

THEOREM 5.6. *Let $f(z)$ be analytic and bounded in $|z| < 1$, let $e^{i\theta_0}$ be an arbitrary point on $K: |z| = 1$, and let E be an arbitrary set of measure zero on K. Then*

$$\varlimsup_{z \to e^{i\theta_0}} |f(z)| \leqslant \varlimsup_{\substack{\theta \to \theta_0 \\ e^{i\theta} \in K - E}} (\varlimsup_{z \to e^{i\theta}} |f(z)|). \tag{5.2.5}$$

Let E_1 denote the set of measure zero on K for which

$$f(e^{i\theta}) = \lim_{r \to 1} f(r e^{i\theta})$$

fails to exist. If $E_z = E \cup E_1$, then

$$\varlimsup_{\substack{\theta \to \theta_0 \\ e^{i\theta} \in K-E}} (\varlimsup_{z \to e^{i\theta}} |f(z)|) \geqslant \varlimsup_{\substack{\theta \to \theta_0 \\ e^{i\theta} \in K-E_z}} (\varlimsup_{z \to e^{i\theta}} |f(z)|) \geqslant \varlimsup_{\theta \to \theta_0} (\sup_{e^{i\theta} \in K-E_z} |f(e^{i\theta})|).$$

By Theorem 5.5, $C(f, e^{i\theta_0})$ is contained in the closure of the convex hull of the set of values $f(e^{i\theta})$ for $e^{i\theta} \in K - E_z$, so that

$$\varlimsup_{\theta \to \theta_0} (\sup_{e^{i\theta} \in K-E_z} |f(e^{i\theta})|) \geqslant \varlimsup_{z \to e^{i\theta_0}} |f(z)|,$$

which completes the proof of Theorem 5.6.

Alternatively, Theorem 5.6 may be stated as

THEOREM 5.6.1. *Let $f(z)$ be analytic and bounded in $|z| < 1$, let $e^{i\theta_0}$ be an arbitrary point of K, and let E be a set of measure zero on K. Then*

$$\varlimsup_{z \to e^{i\theta_0}} |f(z)| \leqslant \varlimsup_{\theta \to \theta_0,\; e^{i\theta} \in K-E} (\sup |f(e^{i\theta})|). \qquad (5.2.6)$$

In order to indicate the scope of (5.2.6), as well as to prove a stronger form of Theorem 5.2, we introduce a modification of the notion of the boundary cluster set $C_B(f, e^{i\theta})$, which was defined in § 6, Chapter 4. Let $e^{i\theta_0}$ be an arbitrary point of K, and let E be some set of measure zero on K. We form first the union

$$C_\rho(f, 0 < |\theta - \theta_0| < \eta; E) = \bigcup_{\substack{0 < |\theta - \theta_0| < \eta \\ e^{i\theta} \notin E}} C_\rho(f, e^{i\theta}) \qquad (5.2.7)$$

for a fixed $\eta > 0$, and then define the set $C_{R-E}(f, e^{i\theta_0})$ as the intersection

$$C_{R-E}(f, e^{i\theta_0}) = \bigcap_{\eta > 0} \bar{C}_\rho(f, 0 < |\theta - \theta_0| < \eta; E). \qquad (5.2.8)$$

This is the radial boundary cluster set modulo E.

Whenever E is empty, we shall write $C_R(f, e^{i\theta_0})$ in place of $C_{R-E}(f, e^{i\theta_0})$. Stated simply, $C_{R-E}(f, e^{i\theta_0})$ is similar to the boundary cluster set $C_B(f, e^{i\theta_0})$ except that the radial cluster sets are used in (5.2.7) instead of the full cluster sets in (4.6.1), and, in addition, we have discarded the radial cluster sets on a set of measure zero. For example, if we consider a Blaschke product $b(z)$ such that every point of K is a limit point of zeros, then $C(b, e^{i\theta})$, and therefore $C_B(b, e^{i\theta})$ as well, consists of the disc $|w| \leqslant 1$ for every θ, so that the set $C(b, e^{i\theta}) - C_B(b, e^{i\theta})$ which must always be open or empty by (5.1.3), is empty. On the other hand, if E is the set of $e^{i\theta}$ for which $b(e^{i\theta})$ is either of modulus less than 1 or does not exist, then $C_{R-E}(b, e^{i\theta_0})$ consists of a closed set on $|w| = 1$. That $C(b, e^{i\theta_0}) - C_{R-E}(b, e^{i\theta_0})$ is open, and therefore consists of the set $|w| < 1$, follows from the following theorem, which generalizes Theorem 5.2.

THEOREM 5.7. *If $f(z)$ is meromorphic in $|z| < 1$, and if E is a set of measure zero on $|z| = 1$, then, at every point $e^{i\theta} \in K$,*

$$\text{Fr}\, C(f, e^{i\theta}) \subseteq C_{R-E}(f, e^{i\theta}). \qquad (5.2.9)$$

If Theorem 5.7 were not true, there would be a point $w_0 \in \text{Fr}\, C(f, e^{i\theta})$ not in $C_{R-E}(f, e^{i\theta})$, so that we can find a $\delta > 0$ such that w_0 is at a distance greater than δ from all points of $C_{R-E}(f, e^{i\theta})$. Since w_0 is a frontier point of $C(f, e^{i\theta})$, we can find a second point w_1 not in $C(f, e^{i\theta})$ such that $|w_1 - w_0| < \frac{1}{2}\delta$; this implies that the distance between w_1 and any point of $C_{R-E}(f, e^{i\theta})$ is greater than $\frac{1}{2}\delta$. Now the function $g(z) = [f(z) - w_1]^{-1}$ is bounded in the neighbourhood of $e^{i\theta}$, and it follows from Theorem 5.6 as applied to the bounded function $g(z)$, that the maximum distance from any finite point to $C(g, e^{i\theta})$ is not greater than the maximum distance from that point to $C_{R-E}(g, e^{i\theta})$, whence

$$\frac{1}{|w_0 - w_1|} \leqslant \frac{2}{\delta},$$

which contradicts our choice of w_1. Hence Theorem 5.7 is established.

If E is any set of measure zero on $|z| = 1$, and if we define the boundary cluster set $C_{B-E}(f, e^{i\theta_0})$ of $f(z)$ at $e^{i\theta_0}$ modulo E as

$$C_{B-E}(f, e^{i\theta_0}) = \bigcap_{\eta > 0} \bar{C}(f, 0 < |\theta - \theta_0| < \eta; E), \qquad (5.2.10)$$

where $\qquad C(f, 0 < |\theta - \theta_0| < \eta; E) = \bigcup_{\substack{0 < |\theta - \theta_0| < \eta \\ e^{i\theta} \notin E}} C(f, e^{i\theta}),$

then it follows that

$$C_{R-E}(f, e^{i\theta_0}) \subseteq C_{B-E}(f, e^{i\theta_0})$$

and $\qquad\qquad \text{Fr}\, C(f, e^{i\theta_0}) \subseteq C_{B-E}(f, e^{i\theta_0}) \qquad\qquad (5.2.11)$

for any point $e^{i\theta_0}$. Thus we have

COROLLARY 3. *Under the hypotheses of Theorem 5.7, the sets*

$$C(f, e^{i\theta}) - C_B(f, e^{i\theta}), \quad C(f, e^{i\theta}) - C_{B-E}(f, e^{i\theta})$$

and $\qquad\qquad C(f, e^{i\theta}) - C_{R-E}(f, e^{i\theta})$

are all open.

Before proceeding to discuss some of the other important consequences of (5.2.6), we should like to discuss the significance in the theory of functions of such results as Theorem 5.2, Theorem 5.7, and Corollary 3, even though these applications will be proved later when we shall demonstrate an extremely

strong and useful generalization† of Picard's theorem to the effect that, *under the hypotheses of Theorem 5.2, every value of*

$$C(f, e^{i\theta}) - C_B(f, e^{i\theta})$$

is assumed infinitely often—with two possible exceptions—in a neighbourhood of a point $e^{i\theta}$ on K, and, if there are any exceptions, these are asymptotic values of $f(z)$ at $e^{i\theta}$. Two examples will serve to illustrate this theorem of Gross and Iversen and to indicate the need of an extension of the form (5.2.9).

The function $w = f(z) = e^{(z+1)/(z-1)}$ is bounded and analytic in $|z| < 1$ and has radial limit of modulus 1 at every point of $|z| = 1$ except at $z = 1$, where $f(z)$ has the radial limit 0. It follows from the classical form of Picard's theorem that every value of $|w| < 1$, except $w = 0$, is assumed infinitely often in $|z| < 1$ in the vicinity of $z = 1$. For $f(z)$ we have $C(f, 1) = \{w : |w| \leqslant 1\}$ and $C_B(f, 1) = \{w : |w| = 1\}$. Now Theorem 5.2 shows that

$$C(f, 1) - C_B(f, 1)$$

consists of the open disc $|w| < 1$, while the theorem of Gross and Iversen shows that every value of $|w| < 1$ is assumed infinitely often, with one possible exception, here $w = 0$, in any neighbourhood of $z = 1$; that $w = 0$ is an asymptotic value (and hence a radial limit value, since $f(z)$ is bounded) is verified by examining the function. At all other points of $|z| = 1$, $C(f, e^{i\theta})$ and $C_B(f, e^{i\theta})$ are degenerate and equal, so that the difference of the two sets is empty.

As a second example, we consider again a Blaschke product $b(z)$ with the property that every point of $K : |z| = 1$ is a limit point of zeros of $b(z)$. Since $C(b, e^{i\theta})$ is the set $|w| \leqslant 1$ for all $e^{i\theta}$, it follows that not only $C(b, e^{i\theta}) - C_B(b, e^{i\theta})$,

but also $C(b, e^{i\theta}) - C_{B-E}(b, e^{i\theta})$,

is empty for every choice of E. On the other hand, if E is chosen so that the radial limit of the modulus of $b(z)$ is 1 everywhere on $K - E$, then $C_{R-E}(b, e^{i\theta})$ is the set $|w| = 1$, and

$$C(b, e^{i\theta}) - C_{R-E}(b, e^{i\theta})$$

is the non-empty set $|w| < 1$. We shall prove in the next section the theorem of Gross and Iversen mentioned above and discuss the extent to which that theorem may be extended.

† Gross [1, 2]; Iversen [1, 2].

3. The Gross–Iversen theorem on exceptional values. Generalizations

We prove first the theorem of Gross and Iversen (Gross [1, 2]; Iversen [1, 2]) referred to in the previous section.

THEOREM 5.8. *If $f(z)$ is meromorphic in $|z| < 1$, then for each point $e^{i\theta}$ on K, every value of $C(f, e^{i\theta}) - C_B(f, e^{i\theta})$ is assumed infinitely often in every neighbourhood of $e^{i\theta}$ with two possible exceptions. Furthermore, every such exceptional value is an asymptotic value of $f(z)$ at $e^{i\theta}$.*

We suppose that there exists a point $P = e^{i\theta_0}$ and a value $\rho_0 > 0$ such that $f(z)$ does not assume in $V_{\rho_0}: |z - P| < \rho_0$ some three values, which we may suppose to be 0, 1 and ∞, belonging to $C(f, e^{i\theta_0}) - C_B(f, e^{i\theta_0})$. We can find numbers $\epsilon_0 > 0$ and $\eta_0 > 0$ such that

$$\liminf_{r \to 1} |f(r e^{i\theta})| > \epsilon_0 \tag{5.3.1}$$

and

$$\liminf_{r \to 1} \frac{1}{|f(r e^{i\theta})|} > \epsilon_0 \tag{5.3.2}$$

for all θ in $0 < |\theta - \theta_0| < \eta_0$; for if, say, (5.3.1) did not hold, there would exist a sequence $\{\theta_k\}$ with $\lim \theta_k = \theta_0$ such that

$$\lim_{k \to \infty} (\liminf_{r \to 1} |f(r e^{i\theta_k})|) = 0,$$

so that 0 would then belong to $C_B(f, e^{i\theta_0})$, which contradicts our assumption that $0 \in C(f, e^{i\theta_0}) - C_B(f, e^{i\theta_0})$.

We assert that there exists an integer n such that at least one component of the open set

$$H_n = \left\{ z : |z| < 1; \ |f| < \frac{1}{n} < \epsilon_0 \right\} \tag{5.3.3}$$

lies completely in V_{ρ_0}. If not, then we may join a point of $V_{\rho_0/2}$ and a point of $|z - e^{i\theta_0}| = \rho_0$ by a continuum κ_1 of FR H_{n+1}, a point of $V_{\rho_0/2^2}$ and a point of $|z - e^{i\theta_0}| = \rho_0$ by a continuum κ_2 of FR H_{n+2}, and, more generally, a point of $V_{\rho_0/2^p}$ and a point of $|z - e^{i\theta_0}| = \rho_0$ by a continuum κ_p of FR H_{n+p}. The sequence $\{\kappa_p\}$ contains a subsequence which converges to a continuum κ_0 joining $e^{i\theta_0}$ to a point on $|z - e^{i\theta_0}| = \rho_0$. No point of κ_0 can lie in $|z| < 1$, since

$f(z) \neq 0$ in V_{ρ_0}; therefore, κ_0 must be an arc of $V_{\rho_0} \cap K$, and the fact that $f(z) \to 0$ as z tends to points of κ_0 violates (5.3.1), so that our assertion follows. If we denote by G a component of H_n lying entirely inside V_{ρ_0}, it follows easily from the maximum modulus theorem that G is simply connected and that the intersection of $\text{Fr}\, G$ with K is precisely the point $e^{i\theta_0}$.

We assert next that G contains at least one component of H_{n+1}. For if $G \cap H_{n+1}$ is empty, we must have in G that

$$\frac{1}{n+1} \leqslant |f(z)| < \frac{1}{n}, \qquad (5.3.4)$$

and, since $|f(z)| = 1/n$ at all points of $\text{Fr}\, G$, except possibly at $e^{i\theta_0}$, an obvious modification of Theorem 5.3 shows that $f(z)$ is a constant of modulus $1/n$. Consequently G contains a nesting sequence of components of H_{n+p} which we may write as

$$G \supset G_1 \supset G_2 \supset \ldots \supset G_p \supset \ldots, \qquad (5.3.5)$$

and which is such that $|f(z)| < 1/(n+p)$ in G_p. It is clear that the diameter of G_p tends to zero as $p \to \infty$, so that we may construct a polygonal path γ_1 terminating at $e^{i\theta_0}$ and passing through the nested sequence (5.3.5) in such a way that $f(z) \to 0$ as $z \to e^{i\theta_0}$ along γ_1. By applying the same argument to the function $1/(f(z))$, we may construct a second path γ_2 terminating at $e^{i\theta_0}$ such that $f(z) \to \infty$ as $z \to e^{i\theta_0}$ along γ_2. Thus we have contradicted Theorem 2.21 of Chapter 2, so that there cannot exist three values in $C(f, e^{i\theta_0}) - C_B(f, e^{i\theta_0})$ which are not assumed infinitely often by $f(z)$ in some neighbourhood of $e^{i\theta_0}$.

To prove the last part of Theorem 5.8, we observe that we have already shown that, if there exist two exceptional values, these values are asymptotic values. Therefore, we may assume that there is only one exceptional value, and that this value is 0. As before we can find numbers $\epsilon_0 > 0$ and $\eta_0 > 0$ for which (5.3.1) holds; furthermore, the same argument shows that at least one component G of the set (5.3.3) lies wholly in V_{ρ_0}, although such a component need not be simply connected. However, the fact that $\bar{G} \cap K$ consists of a single point $e^{i\theta_0}$ together with a modification of Theorem 5.3 enables us to assert the nesting property (5.3.5), so that we may construct, as before, a path γ terminating at $e^{i\theta_0}$ such that $f(z) \to 0$ as $z \to e^{i\theta_0}$ on γ.

Theorem 5.8 contains a natural limitation in view of the example given at the end of §2; indeed, if $b(z)$ is the Blaschke product whose zeros cluster to each point of $|z| = 1$, then, because $C(b, e^{i\theta})$ is the closed disc $|w| \leqslant 1$ for each $e^{i\theta}$, it follows that $C_B(b, e^{i\theta})$ is the closed disc $|w| \leqslant 1$ for each $e^{i\theta}$, so that

$$C(b, e^{i\theta}) - C_B(b, e^{i\theta})$$

is empty for each $e^{i\theta}$. Consequently, we may say that Theorem 5.8 is useful only when the singularities of $f(z)$ on $|z| = 1$ are isolated in some sense. For the function $b(z)$ we know that the radial limits exist with modulus 1 for all $e^{i\theta}$ on $|z| = 1$ with the exception of a possible set E of measure zero. From Theorem 5.7 it follows that $C_{R-E}(b, e^{i\theta})$ consists of the set $|w| = 1$ and that the open set $C(b, e^{i\theta}) - C_{R-E}(b, e^{i\theta})$ is the set $|w| < 1$ for all $e^{i\theta}$. We proceed next to prove a theorem† which is meaningful for a wider class of functions than Theorem 5.8 in that $C(f, e^{i\theta}) - C_{R-E}(f, e^{i\theta})$ need not be empty even when every point of $|z| = 1$ is a singularity of $f(z)$. We have

THEOREM 5.9. *If $f(z)$ is meromorphic in $|z| < 1$, and if E is an arbitrary set of measure zero on $|z| = 1$, then every value of $C(f, e^{i\theta}) - C_{R-E}(f, e^{i\theta})$ is assumed by $f(z)$ in any neighbourhood of $e^{i\theta}$, with the possible exception of a set of capacity zero.*

We may suppose that there exists a point $P = e^{i\theta_0}$ on $K: |z| = 1$ and a closed set $\mathscr{E} \subseteq C(f, P) - C_{R-E}(f, P)$ of positive capacity such that no point of \mathscr{E} is assumed infinitely often in some neighbourhood of P. We may assume that $w = 0$ and $w = \infty$ belong to \mathscr{E} and that that part of \mathscr{E} lying in the circle $|w| < \delta$ is of positive capacity for every $\delta > 0$. We can find $\rho_0 > 0$ such that $f(z)$ is different from 0 and ∞ in $V_{\rho_0}: |z - P| < \rho_0$; furthermore, we can find $\eta_0 > 0$ and $\epsilon_0 > 0$ such that

$$\liminf_{r \to 1} |f(r e^{i\theta})| > \epsilon_0$$

for all θ in $0 < |\theta - \theta_0| < \eta_0$ not belonging to E, for otherwise there would exist a sequence $\{\theta_k\}$ with $\lim \theta_k = \theta_0$ such that

$$\lim_{k \to \infty} (\liminf_{r \to 1} |f(r e^{i\theta_k})|) = 0.$$

† Lohwater [1, 6, 10]; Noshiro [3].

This would then imply that

$$0 \in C_\rho(f, 0 < |\theta - \theta_0| < \eta; E) \quad \text{for every} \quad \eta < \eta_0,$$

so that $w = 0$ would then belong to $C_{R-E}(f, e^{i\theta_0})$ and not to $C(f, e^{i\theta_0}) - C_{R-E}(f, e^{i\theta_0})$, as we assumed.

Now, having chosen ρ_0 and η_0 so that $f(z) \neq 0$ and $f(z) \neq \infty$ in V_{ρ_0} and

$$\liminf_{r \to 1} |f(r\,e^{i\theta})| > \epsilon_0 > 0$$

for all $e^{i\theta}$ on $V_{\rho_0} \cap (K - E)$, we consider the open set

$$H_n = \left\{ z \colon |z| < 1; \ |f(z)| < \frac{1}{n} \right\}$$

for $1/n < \epsilon_0$. We show that *there exists a value of n for which at least one component G of H_n is contained completely in V_{ρ_0}.* Since $0 \in C(f, P)$, there exist points of H_n in every neighbourhood of P, so that if no component of H_n lies wholly in V_{ρ_0} for any $n > 1/\epsilon_0$, we can then find a point in $V_{\rho_0/2}$ and a point of $|z - P| = \rho_0$ which are joined by a continuum κ_1 of $\text{Fr}\,H_{n+1}$, a point in $V_{\rho_0/2^2}$ and a point of $|z - P| = \rho_0$ which are joined by a continuum κ_2 of $\text{Fr}\,H_{n+2}$, and, more generally, a point in $V_{\rho_0/2^p}$ and a point of $|z - P| = \rho_0$ which are joined by a continuum κ_p of $\text{Fr}\,H_{n+p}$. Thus the sequence $\{\kappa_p\}$ must contain a subsequence $\{\kappa_{p_j}\}$ which tends to a limiting continuum κ_0 joining P to a point of $|z - P| = \rho_0$. Since $|f(z)| = 1/(n+p)$ on κ_{n+p}, it is clear that no point of κ_0 can lie in $|z| < 1$; otherwise, $f(z)$ would be identically zero in $|z| < 1$. Consequently, κ_0 must be an arc of K containing P, and every radius drawn to an interior point of κ_0 must cross infinitely many of the κ_{p_j}, so that $0 \in C_\rho(f, e^{i\theta})$ for every $e^{i\theta} \epsilon \kappa_0$, thus contradicting our assumption that $0 \in C(f, P) - C_{R-E}(f, P)$. Hence there exists at least one integer $n > 1/\epsilon_0$ such that at least one component G of H_n lies entirely in V_{ρ_0}.

Since $f(z) \neq 0$ and $f(z) \neq \infty$ in G, it follows at once from the maximum modulus principle that G is simply connected and that the intersection $M = \bar{G} \cap K$ is not empty. We show next that the frontier $\text{Fr}\,G$ of G is locally connected.† Now $\text{Fr}\,G$ is locally con-

† A continuum \mathcal{K} is called locally connected at a point $P_0 \epsilon \mathcal{K}$ if every neighbourhood $U_1(P_0)$ of P_0 contains a second neighbourhood $U_2(P_0)$ of P_0 such that $\mathcal{K} \cap U_2(P_0)$ is connected. The properties of locally connected sets which

nected at each point in $|z| < 1$, for that part of $\mathrm{Fr}\, G$ inside
$|z| < 1$ consists of level curves of the harmonic function $\log |f(z)|$.
Thus if $p \in M$ with $|p| = 1$, and if $\mathrm{Fr}\, G$ is not locally connected
at p, it follows from the definition of local connectedness† that
there exists a non-degenerate subcontinuum H of $\mathrm{Fr}\, G$ con-
taining p such that $\mathrm{Fr}\, G$ is not locally connected at any point of
H; it is clear that H must be an arc of K. There must also exist‡
a circular neighbourhood $V(p)$ and a sequence of mutually dis-
joint components N_1, N_2, \ldots of $\mathrm{Fr}\, G \cap \overline{V(p)}$ converging to a non-
degenerate limiting arc $N \subset H$ containing p. Every radius drawn
to an interior point q of N must cross infinitely many of the com-
ponents N_j $(j = 1, 2, \ldots)$, and on such radii, we have

$$\liminf_{r \to 1} |f(r\, e^{i\theta})| \leqslant \frac{1}{n} < \epsilon_0,$$

which contradicts our hypothesis that $\liminf\limits_{r \to 1} |f(r\, e^{i\theta})| > \epsilon_0 > 0$
almost everywhere. Consequently, $\mathrm{Fr}\, G$ must be locally con-
nected at each of its points.

To prove that the set $M = \mathrm{Fr}\, G \cap K$ has linear measure zero,
we write $M = M_1 \cup (M - M_1)$, where M_1 denotes the subset of M
for which
$$\liminf_{r \to 1} |f(r\, e^{i\theta})| \geqslant \epsilon_0 > 0;$$

clearly, the measure of $M - M_1$ is zero, so that it suffices to show
that $m(M_1) = 0$. We denote by $z = z(\zeta)$ a schlicht function which
maps $|\zeta| < 1$ conformally onto G. Since $\mathrm{Fr}\, G$ is locally connected,
each point of M_1 is arc-wise accessible from G. Let M_1' denote the
set of points on $|\zeta| = 1$ which are mapped into M_1 by $z(\zeta)$. Let p be
an arbitrary point of M_1, and let p_α denote those points of M_1'
corresponding to p under $z(\zeta)$, where $\alpha \in A(p)$, $A(p)$ being some
indexing set. To each point p_α on $|\zeta| = 1$ we draw the radial
segment ρ_α from the circle $|\zeta| = \frac{1}{2}$. Under $z(\zeta)$ these segments ρ_α

we shall use below follow rather easily from the definition, although references
are given for the sake of completeness. The important intuitive notion which
we are using is simply the fact that a continuum cannot fail to be locally con-
nected at an isolated point; rather, the failure must take place on a non-
degenerate subcontinuum.

† See, for example, Whyburn [1], p. 19.
‡ Whyburn [1], p. 18.

are mapped into a family of arcs σ_α, $\alpha \in A(p)$, each of which lies in G except for the one end-point p, and which are mutually disjoint except for the common end-point p. Furthermore, since $p \in M_1$, there exists a positive number $\delta(p) > 0$ which represents the distance from p to the nearest point of $\mathrm{F_R}\, G$ along the radius of $|z| < 1$ drawn to p. Letting $e^{i\theta_p}$ represent the point p, we form the radial segment

$$r_p = \{r\, e^{i\theta}: \theta = \theta_p; 1 - \tfrac{1}{2}\delta(p) \leqslant r \leqslant 1 + \tfrac{1}{2}\delta(p)\},$$

and consider the set of points

$$T_p = r_p \cup \sigma_p \quad (p \in M_1).$$

The set T_p contains at least one plane triod† with junction point p. By Moore's theorem,† the set M_1 is at most countable, so that M has linear measure zero.‡

If $z = z(\zeta)$ is the function mapping $|\zeta| < 1$ conformally onto G, then the function $F(\zeta) = nf(z(\zeta))$ is analytic and bounded, $|F(\zeta)| < 1$, in $|\zeta| < 1$. By Theorem 3.4, Chapter 3, the image M_ζ of M under $z(\zeta)$ is of linear measure zero, so that the function $F(\zeta)$ has radial limits of modulus 1 almost everywhere on $|\zeta| = 1$. By Theorem 2.14, Chapter 2, a function $w = F(\zeta)$ with these properties, which is neither a constant nor a finite Blaschke product, must assume infinitely often in $|\zeta| < 1$ all values of $|w| < 1$ except for a possible set of capacity zero. Since $F(\zeta) \neq 0$ in $|\zeta| < 1$, $F(\zeta)$ cannot be a finite Blaschke product, so that $f(z)$ must assume infinitely often in G all values in $|w| < 1/n$ except possibly a set of capacity zero; we assume, of course, that $f(z)$, and a $fortiori$ $F(\zeta)$, is not constant. Since we may take ρ_0 arbitrarily small, we have forced the contradiction which proves Theorem 5.9.

In the proof of Theorem 5.9 we were able to show that there exists a value of n for which at least one component of H_n is completely contained in V_{ρ_0}. In such a component G_n there exists at

† A *plane triod* T is the union of two arcs h_1 and h_2 whose intersection is a single point P which is simultaneously an end-point of h_1 and an interior point of h_2; the point P is called the *junction point* of T. A theorem of Moore [1] asserts that any set of mutually disjoint triods in the plane is countable.

‡ It is clear that the same argument shows that the capacity of M is zero whenever E has capacity zero.

least one component of the set H_{n+1}, for otherwise, we must have in G_n

$$\frac{1}{n+1} < |f(z)| < \frac{1}{n}.$$

Since $|f(z)| = 1/n$ at all points of Fr G_n lying inside $|z| < 1$, and since $K \cap$ Fr G_n is of linear measure zero, it follows that the same sort of argument used in the proof of Theorem 5.8 allows us to assert that there exists a path γ terminating at a point $e^{i\alpha}$ of $V_{\rho_0} \cap K$ such that $f(z) \to 0$ as $z \to e^{i\alpha}$ along γ. Thus we have

THEOREM 5.10. *Every exceptional value of $f(z)$ in*

$$C(f, P) - C_{R-E}(f, P)$$

is an asymptotic value of $f(z)$, either at P, or else at a sequence of points $e^{i\alpha_k}$ having P as a limit point.

The example given in Chapter 1, § 8, shows that Theorem 5.9 is best possible in that there exists a function $w = f(z)$ with the property that the set $C(f, P) - C_{R-E}(f, P)$ is precisely the open circle $|w| < 1$ and such that $f(z)$ omits all values belonging to a given closed set E_w of capacity zero lying in $|w| < 1$.

4. Properties of functions of Seidel's class U

We shall apply Theorems 5.9 and 5.10 to various of the classes of functions discussed earlier. We remark first that Theorem 2.14 is a special case of Theorem 5.9. Indeed, the class of functions possessing almost everywhere on an arc radial limits of a constant modulus, or more generally, the class of functions the radial limit of whose modulus is constant almost everywhere on some arc of $|z| = 1$, can be treated as a special case of the class of functions described by Theorems 5.4 and 5.10. We have first†

THEOREM 5.11. *Let $f(z)$ be analytic and bounded in $|z| < 1$, and let the radial limits of $f(z)$ have modulus 1 almost everywhere on $|z| = 1$. If $f(z)$ omits a value w_0 with $|w_0| < 1$, then, unless $f(z)$ is identically constant in $|z| < 1$, there exists at least one radius $\theta = \theta_0$ such that $\lim_{r \to 1} f(r e^{i\theta_0}) = w_0$.*

† Frostman [1], Hössjer and Frostman [1], Seidel [2]. For a different proof, cf. Lohwater [5].

Theorem 5.11 has been extended† by showing that the condition of boundedness may be dispensed with; it is clear that Lindelöf's theorem need not be applicable in the case of an unbounded function, so that the radial approach in Theorem 5.11 must, in general, be sacrificed. We have

THEOREM 5.12. *Let $f(z)$ be analytic in $|z| < 1$, and let the modulus $|f(r\,e^{i\theta})|$ have radial limit 1 for almost all $e^{i\theta}$ on $|z| = 1$. If $f(z)$ omits the value 0 in $|z| < 1$, then, unless $f(z)$ reduces to a constant in $|z| < 1$, $f(z)$ must admit either $w = 0$ or $w = \infty$ as an asymptotic value. If $w = \infty$ is not an asymptotic value of $f(z)$, then $|f(z)| < 1$ in $|z| < 1$.*

Suppose next that $f(z)$ is analytic and bounded, $|f(z)| < 1$, in $|z| < 1$, and that the radial limits of $f(z)$ are of modulus 1 almost everywhere on an arc $A: (\theta_1 < \theta < \theta_2)$ of $|z| = 1$. We assume that P is a singularity of $f(z)$ on A, that is, that the reflexion principle of Theorem 5.4 is not valid on any arc containing P. If P is an isolated singularity, then

$$C(f,P) - C_B(f,P) = C(f,P) - C_{R-E}(f,P) = \{w: |w| < 1\},$$

and any exceptional value of $f(z)$ is an asymptotic value, and, since $f(z)$ is bounded, a radial limit value of $f(z)$. Hence $f(z)$ can omit at most one value of $C(f,P) - C_B(f,P)$ in any neighbourhood of P, so that if a bounded function omits two (or more) values of $C(f,P) - C_{R-E}(f,P)$ in the neighbourhood of a singularity P, the singularity cannot be isolated. Since the set of singularities on A is closed, we have

THEOREM 5.13.‡ *If $f(z)$ is analytic and bounded, $|f(z)| < 1$, with radial limits of modulus 1 almost everywhere on an arc $A(\theta_1 < \theta < \theta_2)$ of $|z| = 1$, and if $f(z)$ omits two values of modulus less than 1 in the vicinity of A, then the set of singularities of $f(z)$ on A is perfect.*

For various other properties of a function whose modulus has a fixed radial limit almost everywhere, the reader is referred to the bibliography§ of recently proved results. We mention without proof an interesting analogue of Frostman's theorem (Theorem 2.14), which is concerned with the problem of reducing

† Lohwater [1, 5]. ‡ Seidel [2]. § Lohwater [1, 3–9].

the size of the exceptional set appearing in Theorem 5.9. The analogue in question is

THEOREM 5.14.† *Let* $w = f(z)$ *be analytic and bounded,* $|f(z)| < 1$, *in* $|z| < 1$, *and let the radial limit of the modulus* $|f(r\,e^{i\theta})|$ *be* 1 *for all* $e^{i\theta}$ *on* $|z| = 1$, *except possibly for a set of capacity zero on* $|z| = 1$. *Then, unless* $f(z)$ *reduces to a finite Blaschke product or to a constant, every value of* $|w| < 1$ *is assumed infinitely often by* $f(z)$ *with at most one exception.*

By means of this theorem we may prove

THEOREM 5.15. *If, under the hypotheses of Theorem 5.9, the set E is of capacity zero, the exceptional set is at most countable.*

The proof of Theorem 5.15 is a direct modification of that of Theorem 5.9, except that Theorem 5.14, the proof of which is beyond the scope of this monograph, is used in place of Frostman's theorem. Using methods which are again beyond the scope of this monograph, Woolf [1, 2] has proved an interesting extension of Theorem 5.15, namely,

THEOREM 5.16. *If, under the hypotheses of Theorem 5.9, the set E is of capacity zero, then* $f(z)$ *assumes every value of*

$$C(f, e^{i\theta}) - C_{R-E}(f, e^{i\theta})$$

in any neighbourhood of $e^{i\theta}$, *with the possible exception of two values in each component of* $C(f, e^{i\theta}) - C_{R-E}(f, e^{i\theta})$.

Using methods involving Banach algebras, Weiss [1] has shown recently that these theorems are best possible.

5. Scope of the foregoing theorems

We complete this chapter with some remarks on the application of these theorems to the question of the distribution of values of a meromorphic function in the vicinity of its singularities on $|z| = 1$. As the example of the Blaschke product $b(z)$ showed, Theorem 5.8 is empty whenever the set $C(f, P) - C_B(f, P)$ is empty; indeed, since $C(b, e^{i\theta})$ is the closed disc $|w| \leqslant 1$ for every $e^{i\theta}$, it follows that every boundary cluster set formed by means of $C(b, e^{i\theta})$ will coincide with $C(b, e^{i\theta_0})$ no matter what the size of the set E which is discarded. Thus, for any set E on $|z| = 1$ which is

† Lohwater [7].

dense in the neighbourhood of P the set $C_{B-E}(b, e^{i\theta_0})$ is the set $|w| \leqslant 1$, so that $C(b, e^{i\theta_0}) - C_{B-E}(b, e^{i\theta_0})$ is empty. Therefore, the hopes of finding meaningful generalizations of Theorem 5.8 to the case of non-isolated singularities rest with the formation of boundary cluster sets which are defined by means of proper subsets of the cluster sets $C(f, e^{i\theta})$. Consequently, we are forced, in Theorems 5.9 and 5.15, to define the set $C_{R-E}(f, e^{i\theta_0})$ by means of the radial cluster sets $C_\rho(f, 0 < |\theta - \theta_0| < \eta; -E)$. It is interesting to note, however, that Theorems 4.8 and 4.10 of Chapter 4 show that there exists a certain limitation on the extent to which we can form boundary cluster sets, for, given an arbitrary continuum γ_0 in $|z| \leqslant 1$ intersecting $|z| = 1$ at the point $z = 1$, we have $C(f, e^{i\theta}) = C_{\gamma_\theta}(f, e^{i\theta})$ for a residual set on $|z| = 1$, where γ_θ denotes the rotation of γ_0 through an angle θ about $z = 0$. Thus we have

THEOREM 5.17. *For all $e^{i\theta}$ on $|z| = 1$, except for a countable set, the set $C(f, e^{i\theta}) - C_B(f, e^{i\theta})$ is empty, while, except for a set of first category, the sets $C(f, e^{i\theta}) - C_{B-E}(f, e^{i\theta})$, and $C(f, e^{i\theta}) - C_{R-E}(f, e^{i\theta})$ are empty unless possibly E is residual on $|z| = 1$.*

Theorem 4.8 of Chapter 4 also shows that we may not expect to improve Theorem 5.9 by defining new boundary cluster sets by means of continua other than radii in the formation of proper subsets of $C(f, e^{i\theta})$. It is clear that analogues of Theorem 5.9 and Theorem 5.15 may be formulated using continua other than radii in the formation of the boundary cluster set.†

If, on the other hand, the class of functions can be further restricted so that the radial limits of the modulus are constant everywhere, except for a set E which is either of measure zero or capacity zero, as is the case in Theorems 5.11 through 5.14, there is no problem concerning the possible empty character of $C(f, P) - C_{R-E}(f, P)$.

6. One-sided cluster-set theorems

A recent theorem of Doob [13] belongs to the same order of ideas as Theorems 5.2 and 5.3, only the relations established are one-sided.

† See, for example, Collingwood and Cartwright [1], Theorem 19, where the radii are replaced by asymptotic paths whose ends converge to the points $e^{i\theta}$.

We require the following definitions. If $0 < \omega < \pi$ the subset $\{z \colon |z| < 1, 3\pi/2 - \arg(z - e^{i\theta}) \leqslant \omega - \theta\}$ of the unit disc will be called the right ω-sector at the point $z = e^{i\theta}$, with a similar definition for the left ω-sector, and the bounding chord will be called the right (or left) ω-chord.

Let $C_\omega(f, e^{i\theta})$ be the cluster set of f in the right (left) ω-sector, and let

$$C^\omega(f, e^{i\theta}) = \bigcup_{\omega' < \omega} C_{\omega'}(f, e^{i\theta});$$

$$\mathcal{T}_0(f, e^{i\theta}) = \bigcap_{\omega > 0} C^\omega(f, e^{i\theta}) = \bigcap_{\omega > 0} C_\omega(f, e^{i\theta}).$$

We call $\mathcal{T}_0(f, e^{i\theta})$ the *right (left) tangential cluster set* of f at $e^{i\theta}$; both \mathcal{T}_0 and C_ω are closed.

The inequality $\omega' < \omega$ implies that $C^\omega \supset C_{\omega'} \supset C^{\omega'}$, which shows that C^ω contains the closure of $C^{\omega'}$.

The theorem to be proved is

THEOREM 5.18. *If $f(z)$ is meromorphic in $|z| < 1$, then at every point $e^{i\theta}$ of $|z| = 1$,*

(i) $\mathrm{FR}\, \mathcal{T}_0(f, e^{i\theta}) \subseteq C_{Br}(f, e^{i\theta})$,

where $\mathrm{FR}\, \mathcal{T}_0$ denotes the frontier of the right tangential cluster set, with a similar inclusion by C_{Bl} for the left tangential cluster set;

(ii) *if $\omega > 0$,* $C^\omega(f, e^{i\theta}) \cap \mathscr{C} C_{Br}(f, e^{i\theta})$

is open (or empty) and each open component of this set has a frontier point on $C_{Br}(f, e^{i\theta})$.

There are similar assertions for the left-hand sets. For the proof we take $\theta = 0$ and consider the right sectorial, tangential and boundary cluster sets at the point $z = 1$. We need the following

LEMMA. *Let f be a bounded regular function in the unit disc and let A be an arc of $|z| = 1$ having $z = 1$ as an end-point and lying to the right (left) of this point. Then if $\alpha \in \mathcal{T}_0(f, 1)$ we have $|\alpha| \leqslant |\beta|$ for all $\beta \in C_{Br}(f, 1)$.*

Consider a segment $S(A_1)$ of the unit disc cut off by a chord Q to $z = 1$ and the arc A_1 abutting this point on the right. We may assume $|f| < 1$ in the unit disc and $|\gamma| < k < 1$ for all $\gamma \in C(f, P)$, $P \in A_1$. Then, given $\epsilon > 0$, $|f| < k + \epsilon$ at all points of $S(A_1)$ sufficiently near to A_1. Now reflect $S(A_1)$ in Q and rotate the domain

bounded by A_1 and its image so that Q is in the real axis, and denote the resulting domain by $S'(A_1)$ and the transform of f by $\varphi(\xi)$. The function

$$\Phi(\xi) = \varphi(\xi)\,\bar{\varphi}(\bar{\xi})$$

is regular in $S'(A_1)$ and satisfies $\Phi(\xi) < k + \epsilon$ at all points of $S'(A_1)$ sufficiently near to its boundary. Hence, by the maximum modulus principle,

$$|\Phi(\xi)| = |\varphi(\xi)|^2 \leqslant k$$

for ξ on the real axis and thus $|f| \leqslant k^{\frac{1}{2}}$ on the chord bounding $S(A_1)$. Therefore $|f| \leqslant k^{\frac{1}{2}}$ throughout $S(A_1)$.

A circle C of unit radius whose centre lies on the exterior normal to the bounding chord of $S(A_1)$ through $z = 1$ cuts off an arc $A_2 \subset A_1$ which defines a new domain $S(A_2)$ whose image in the chord across A_2 is the segment of C intercepted by A_2 and thus lies in $S(A_1)$ so that in the domain consisting of $S(A_2)$ and its image $|f| \leqslant k^{\frac{1}{2}}$ while, again, $|f| < k + \epsilon$ sufficiently near to A_2. Repeating the previous argument we obtain

$$|f| \leqslant k^{1 - \frac{1}{4}} \text{ in } S(A_2),$$

and after n repetitions

$$|f| \leqslant k^{1 - 1/2^n} \text{ in } S(A_n).$$

By choice of A_1 we may take k as close as we please to the maximum of $|\beta|$ for $\beta \in C_{Br}(f, 1)$. The lemma is therefore proved.

Now, to prove (i) suppose α is a boundary point of $\mathscr{T}_0(f, 1)$ not in $C_{Br}(f, 1)$. Then $\alpha \in \mathscr{T}_0$ since this set is closed. Let 3δ be the distance from α to C_{Br} and let α' not in \mathscr{T}_0 be such that $|\alpha - \alpha'| < \delta$. Then α' cannot be in every $C_\omega(f, 1)$ for $\omega > 0$ and we can choose $\omega' > 0$ such that the function $1/(f - \alpha')$ is bounded in the ω'-sector near $z = 1$. Since the distance from α' to C_{Br} is greater than 2δ it follows that $|\beta| \leqslant 1/2\delta$ for all $\beta \in C_{Br}(1/(f - \alpha'), 1)$ so that, by the lemma, the right tangential cluster values of $1/(f - \alpha')$ have modulus $1/2\delta$ which contradicts the choice of α'.

To prove (ii) suppose α is a frontier point of $C_\omega(f, 1)$ not in $C_{Br}(f, 1)$. There is a sequence $\{z_n\}$ in some ω'-sector, $\omega' < \omega$, with limit 1 on which $f(z_n) \to \alpha$. Let L_n be the linear transformation taking the unit disc onto itself, 1 into 1, 0 into z_n, and define $f_n(z) = f(L_n(z))$. If $\{f_n\}$ is a normal family, any limit function must be identically α, or else there is a closed disc about the origin small

enough to be carried by L_n into the interior of the ω-sector for all
large n and on which the values of $f_n(z)$ cover a neighbourhood
of α, which neighbourhood would lie in C_ω, contrary to hypothesis.
Hence $f_n \to \alpha$ uniformly in any disc $|z| < h < 1$ from which it
follows that $\alpha \in \mathcal{T}_0$. But this is inconsistent with (i). Hence $\{f_n\}$ is
not a normal family. The same reasoning can be applied to the
ω-sector, or a portion of it near to 1, after a preliminary mapping
onto the unit disc. Therefore the closure of the range of the
sequence $\{f_n\}$ covers the sphere so that $C^\omega(f, 1)$ is total, contrary
to the hypothesis on α. This proves the first assertion of (ii). The
second follows since $C^\omega(f, 1)$ is connected.

This theorem, apart from its general interest, is important for
the theory of *principal values* of a function, that is to say values
which are cluster values on every path to a boundary point of the
disc. We shall return to this subject in Chapter 9.

CHAPTER 6

BOUNDARY THEORY IN THE LARGE

1. Method of the inverse function

Important results are obtained by studying the inverse $z = z(w) = f^{-1}(w)$ of the analytic function $w = f(z)$. These results turn for the most part upon the relationship between the singularities of the function $z(w)$ and the set $\Gamma(f)$ of asymptotic values of $f(z)$. We therefore begin this chapter with a brief account of those properties of the inverse function which we shall require.

We write $w = f(z)$. Then, to any z_0 in the domain of existence D of $f(z)$ there corresponds a power series $e_{z_0} = e_{z_0}(w, w_0)$ in $(w - w_0)^{1/k}$ having a positive radius of convergence, where $w_0 = f(z_0)$ and $k \geqslant 1$ according as w_0 is a simple or multiple value of $f(z)$ at z_0. We may assume that $z = \infty$ is not contained in D so that $z_0 \neq \infty$, e_{z_0} contains no negative powers, and w_0 is not a pole. If z_0 is a pole, w_0 becomes the point at infinity and e_{z_0} takes the form of a power series in $(1/w)^{1/k}$. This distinction, awkward in the case of a meromorphic function having an infinite number of poles, is obviated in the usual way by projecting the w-plane stereographically onto the sphere of unit diameter tangent to the plane at the origin—the Riemann sphere. If $k = 1$, the series e_{z_0} maps the interior of its circle of convergence c_{z_0} conformally onto a domain $d(z_0)$ containing z_0 and containing no zero of $f'(z)$, the derivative of $f(z)$. If $k > 1$, the mapping is conformal except at the point z_0, which is a zero of $f'(z)$, and $d(z_0)$ contains precisely k simple zeros of $f(z) - w' = 0$ for every $w' \neq w_0$ contained in c_{z_0}. We call e_{z_0} an *element* of the inverse function $z(w)$ of $w = f(z)$; and if $k = 1$, that is to say, if $f'(z_0) \neq 0$, e_{z_0} is a *regular element*, while if $k > 1$ it is an *algebraic element* with its (algebraic) branch point at $w_0 = f(z_0)$, where $f'(z_0) = 0$.

Now consider two points z_1, z_2 in D, which is a connected open set. z_1 can be joined to z_2 by a simple polygonal path L in D, at all points of which $f(z)$ is meromorphic and $f'(z) \neq 0$, and this path is mapped by $f(z)$ onto a piecewise analytic curve Λ in the w-plane

or on the w-sphere. To every point z' of L there corresponds an inverse element $e_{z'}(w, w')$, $w' = f(z')$, which maps the interior of the circle of convergence $c_{z'}$ of $e_{z'}$ conformally onto a domain $d(z')$ containing z'. L together with its end-points z_1, z_2 forms a closed set every point z' of which is covered by the domain $d(z')$ derived from it by the inverse mapping we have just described. It follows from the Borel covering theorem that L is covered by a finite number of the domains $d(z')$. Since consecutive domains of this finite set overlap, the corresponding circles in the w-plane or on the w-sphere with their centres on Λ also overlap; and the corresponding elements are equal in the common parts of these circles and thus are immediate continuations of one another. Every other element $e_{z'}$ corresponding to a point z' on L is an immediate continuation of one of this finite set of elements and is thus an element of the continuation of $z(w)$ along Λ from $w_1 = f(z_1)$ to $w_2 = f(z_2)$ corresponding to the continuation of $f(z)$ along L from z_1 to z_2.

We have now only to apply the same argument to the elements $e_{w'}(z, z')$ of the function $f(z)$ in D to show that to every continuation of an element $e_{z_1}(w, w_1)$ of $z(w)$ to another element $e_{z_2}(w, w_2)$ there corresponds a continuation of the element $e_{w_1}(z, z_1)$ of $f(z)$, where $w_1 = f(z_1)$, of which $e_{z_1}(w, w_1)$ is the inverse, to an element $e_{w_2}(z, z_2)$, $w_2 = f(z_2)$, of which $e_{z_2}(w, w_2)$ is the inverse.

We have thus proved (i) *that given any two elements $e_{z_1}(w, w_1)$ and $e_{z_2}(w, w_2)$ of $z(w)$ in the neighbourhoods of $z_1 \in D$ and $z_2 \in D$ respectively, each is a continuation of the other*; and (ii) *that given any element $e_{z_2}(w, w_2)$ which is a continuation of an element $e_{z_1}(w, w_1)$ of $z(w)$, then $e_{z_2}(w, w_2)$ is the inverse of an element of $f(z)$.* $z(w)$ is thus defined throughout its domain of existence by analytic continuation from any element inverse to an element of $f(z)$, and the aggregate of the inverse elements $e_{z'}(w, w')$, $z' \in D$, constitutes the function $z(w)$. Except in the case that $f(z)$ is univalent in D, its inverse $z(w)$ is a multiform function whose domain of existence is a Riemann surface which we may take at our convenience as covering either the plane or the Riemann sphere.

Continuation of $z(w)$ through an algebraic point z_0, $f'(z_0) = 0$, involves the inclusion in the continuation of a branch of the corresponding algebraic element e_{z_0}. In speaking of continuations of

$z(w)$, both regular and algebraic elements are allowed in any particular continuation unless expressly stated to the contrary. A continuation through an algebraic point defines a particular branch of $z(w)$ along the path of continuation or, to put the matter in another way, is a continuation along one of a number of paths on the Riemann surface all passing through the branch point in question.

Now consider a path λ in the w-plane (or on the w-sphere) defined by a continuous function $w = w(t)$, $0 \leqslant t < 1$. Then, by definition, an analytic continuation along λ is a set of regular or algebraic elements $e_{z_t}(w', w(t))$, where $z_t = e_{z_t}(w', w(t))$, such that for any t in $0 \leqslant t < 1$ the element $e_{z_t}(w', w(t))$ exists and we can find $\epsilon = \epsilon(t) > 0$ such that for $|T - t| < \epsilon$ all the elements $e_{z_T}(w', w(T))$ are immediate continuations of $e_{z_t}(w', w(t))$. Any two elements of the set can be joined by a finite chain of elements of the set which may be selected in an infinity of different ways. It follows at once from the foregoing argument that to every continuation of an element of $z(w)$ along λ, supposing such a continuation to exist, there corresponds a continuation of an element of $f(z)$ along the corresponding path l in the z-plane. We note expressly that either or both of the paths λ and l may be closed curves described any finite number of times.

Thus far we have placed no restriction on D, which may be any connected open set in the z-plane. We shall now consider special cases. The simplest is the case in which D is simply connected. If the frontier of D contains more than one point, then, by the fundamental theorem of conformal mapping, it can be mapped onto the open unit disc $|z| < 1$. We need therefore only consider the two typical situations in which $f(z)$ is meromorphic in the punctured plane (or sphere) $z \neq \infty$, or in the unit disc $|z| < 1$, as the case may be. Again by the fundamental mapping theorem, the Riemann surface onto which the punctured plane or the open disc is mapped by a meromorphic function for which it is the domain of existence (so that the point $z = \infty$ or the unit circumference $|z| = 1$ is a natural boundary) is simply connected and the mapping is locally conformal except at algebraic points. Also, the inverse mapping by elements of $z(w)$ of the Riemann surface into the punctured plane is, as we have seen, locally conformal except

at zeros of $f'(z)$. Any open simply connected Riemann surface can be mapped in this way either upon the punctured plane or upon the open disc. In the former case it is said to be of *parabolic type* and in the latter of *hyperbolic type*. These types are mutually exclusive.

We have also to consider the case where $f(z)$ is meromorphic in $|z| < 1$ but can be continued across the circumference $|z| = 1$ to the domain $|z| > 1$. To eliminate trivial cases we assume there is at least one essential singularity on the circumference $|z| = 1$ which is not, however, a natural boundary. We therefore classify the elements $e_z(w', w)$ of the inverse function $z(w)$ of $f(z)$ as follows: $e_z(w', w)$ is an *internal element* of $z(w)$ if it is the inverse of an element $e_w(z', z)$ of $f(z)$ for which $|z| < 1$; it is a *boundary element* if $|z| = 1$; and it is an *external element* if $|z| > 1$. If the circumference $|z| = 1$ is a natural boundary, then $z(w)$ has only internal elements.

It can be shown trivially that *any continuation of an element of $z(w)$ along a given continuous path λ defined by $w = w(t)$, $0 \leqslant t < 1$, which contains both internal and external elements contains at least one boundary element*. It follows as a corollary that *any continuation along a path λ in the w-plane (or on the w-sphere) of an element of $z(w)$ which contains no boundary element contains only internal or only external elements*.

Taking $f(z)$ to be meromorphic in the open disc $|z| < 1$, let us now suppose that the circumference $|z| = 1$ is a natural boundary, and let us consider a path λ defined by a continuous function $w = w(t)$, $0 \leqslant t < 1$ such that $\lim_{t \to 1} w(t) = \omega$. Suppose further that there is a continuation of an element of $z(w)$ along λ towards the point ω such that the radii of convergence of the regular or algebraic elements $e_{z_t}(w', w(t))$ tend to zero as $t \to 1$. Then ω is neither a regular nor an algebraic point, nor can it be a pole of the branch of $z(w)$ generated by the continuation. ω is called a *transcendent singularity* of the inverse function $z(w)$. The path λ is mapped by this branch of $z(w)$ onto a path l defined by the continuous function $z = \zeta(t) = z(w(t))$, $0 \leqslant t < 1$, where $|\zeta(t)| < 1$. We see at once that $\lim_{t \to 1} |\zeta(t)| = 1$. For, if not, there is a number $r_0 < 1$ and a sequence $t_1 < t_2 < \ldots < t_n < \ldots$, $\lim_{n \to \infty} t_n = 1$, such

that $\lim_{n\to\infty} |\zeta(t_n)| = r_0$, and hence there is a point $z_0 = r_0 e^{i\theta_0}$ and a subsequence $\{t_\nu\}$ such that $\lim_{\nu\to\infty} \zeta(t_\nu) = z_0$. Now since $|z_0| < 1$ there is a regular or algebraic element $e_{z_0}(w'\,w_0)$, $w_0 = f(z_0)$, of $z(w)$ having a positive radius of convergence ρ_0. We can find ν_0 such that $|\zeta(t_\nu) - z_0| < \frac{1}{2}\rho_0$ for all $\nu > \nu_0$, and it follows that each of the elements $e_{z_t}(w', w(t_\nu))$ is an immediate continuation of $e_{z_0}(w', w_0)$ and thus has a radius of convergence greater than $\frac{1}{2}\rho_0$ in contradiction to the hypothesis that the radius of convergence of $e_{z_t}(w', w(t))$ tends to zero as $t \to 1$. Since $\lim_{t\to 1} f(\zeta(t)) = \omega$ and $\lim_{t\to 1} |\zeta(t)| = 1$, we see that $\omega \in \Gamma(f)$.

Conversely, if $\omega \in \Gamma(f)$ then ω is a transcendent singularity for some branch of $z(w)$. For there is a path l in $|z| < 1$ defined by a continuous function $z = z(t)$, $0 \leqslant t < 1$, such that $\lim_{t\to 1} |z(t)| = 1$ and $\lim_{t\to 1} f(z(t)) = \omega$. We may assume without loss of generality that l passes through no zero of $f'(z)$ so that l is mapped by $f(z)$ on a continuous curve λ, $w = f(z(t))$, $0 \leqslant t < 1$, without branch points such that $f(z(t)) \to \omega$ as $t \to 1$. Let $z_1 = z(t_1) \in l$ and $e_{z_1}(w', w_1)$, $w_1 = f(z_1) \in \lambda$, be the corresponding element of $z(w)$. If the radius of convergence of the elements $e_z(w', w_1)$, obtained by continuation of $e_{z_1}(w', w_1)$ along λ towards ω does not tend to zero, then this radius has a lower bound ρ_0. We can find t_0 such that $|f(z(t)) - \omega| < \frac{1}{2}\rho_0$ for $t > t_0$. The continuation contains the element $e_{z_0}(w', w_0)$, where $z_0 = z(t_0)$, $w_0 = f(z(t_0))$, whose circle of convergence, of radius not less than ρ_0, contains the circle $|w - \omega| = \frac{1}{2}\rho_0$, which in turn contains the curve $f(z(t))$, $t_0 < t < 1$, which it therefore maps within a domain contained in $|z| < 1$. But since the continuation maps λ onto l on which $|z| \to 1$ as $t \to 1$, this gives a contradiction, and it follows that ω must be a transcendent singularity for the branch of $z(w)$ generated by this continuation. We have thus proved

THEOREM 6.1. *If* $|z| = 1$ *is a natural boundary for a function* $f(z)$ *meromorphic in* $|z| < 1$, *then* $\Gamma(f) = \Omega(f)$, *where* $\Omega(f)$ *is the set of transcendent singularities of the inverse* $z(w)$ *of* $w = f(z)$.

Next, let us consider the case that $|z| = 1$ is not a natural boundary for $f(z)$. There are then non-singular points (among

which we include the poles and algebraic points of $f(z)$) on $|z| = 1$. These form an open set which thus consists of a finite or countable set of open intervals I_n on $|z| = 1$. At each point $e^{i\theta}$ of $I = \bigcup_n I_n$, $f(z)$ clearly has a unique asymptotic value, which we call a *regular boundary value*, and which corresponds to a boundary element of $z(w)$. The intervals I_n are mapped by $f(z)$ onto a set of analytic arcs every point of which is a regular asymptotic value; it follows that *the set of regular asymptotic values of $f(z)$ is either empty or of positive linear measure, in which case it contains a continuum.*

Asymptotic values of $f(z)$ which are not regular we call *singular*. If there is a continuation along λ defined by $w = w(t)$, $0 \leqslant t < 1$, $\lim_{t\to 1} w(t) = \omega$, consisting only of internal elements of $z(w)$ and such that the radius of convergence tends to zero as $t \to 1$, then ω is a transcendent singularity for the branch of $z(w)$ generated by the continuation and, as we have shown above, λ is mapped by this branch of $z(w)$ onto a path l defined by $z = \zeta(t)$, $0 \leqslant t < 1$, which cannot have any point of $|z| < 1$ or any regular point of $|z| = 1$ in its limiting set as $t \to 1$. Hence $\lim_{t\to 1} |\zeta(t)| = 1$ and ω *is a singular asymptotic value of $f(z)$.* Conversely, if $\omega \in \Gamma(f)$ and is not a regular asymptotic value then ω is a transcendent singularity for some *internal* branch of $z(w)$; i.e. for some branch consisting only of internal elements.

Denote by $\Omega(f)$ the set of transcendent singularities of the *internal* branch of $z(w)$ consisting of all internal elements† and by $\Upsilon(f)$ the set of regular boundary values. We have proved that *if $|z| = 1$ is not a natural boundary for $f(z)$, then*

$$\Gamma(f) = \Omega(f) \cup \Upsilon(f).$$

Combining this with Theorem 6.1 we have

THEOREM 6.2. *If $f(z)$ is meromorphic in $|z| < 1$, then*

$$\Gamma(f) = \Omega(f) \cup \Upsilon(f),$$

† If $|z| = 1$ is not a natural boundary, the complete function $f(z)$ generated by continuation across the circumference $|z| = 1$ in both directions may not be uniform, so that for any z in $|z| < 1$ there may be more than one element $e_w(z', z)$ of the complete function. But as we are only concerned with a single branch of $f(z)$ which is, by hypothesis, uniform in $|z| < 1$, no ambiguity arises.

where $\Omega(f)$ is the set of transcendent singularities for the internal branch of $z(w)$ and $\Upsilon(f)$ is the set of regular boundary values.

If $|z| = 1$ is a natural boundary, then $\Upsilon(f)$ is empty, and the internal branch is the complete function $z(w)$ so that $\Gamma(f) = \Omega(f)$. Hence we have

COROLLARY. *If $\Gamma(f)$ does not contain a continuum, $\Upsilon(f)$ is empty, so that $|z| = 1$ is a natural boundary for $f(z)$, and $\Gamma(f) = \Omega(f)$.*

Similar considerations apply where D is multiply connected.

2. The set $\Gamma(f)$ for bounded functions

We wish to study $\Gamma(f)$ for $f(z)$ meromorphic in $|z| < 1$, where $\mathscr{C}C(f)$ is not empty. If $a \in \mathscr{C}C(f)$, there are positive numbers σ and ϵ such that $|f(z) - a| > \sigma$ (or $|f(z)| < 1/\sigma$ if $a = \infty$) in the domain $1 - \epsilon < |z| < 1$. If $a \neq \infty$, we make a linear transformation that puts a at ∞ so that the transform of $f(z)$ is bounded. It is therefore sufficient to consider functions regular and bounded in a disc. The method of the inverse function enables us to prove a theorem which is fundamental for our purpose in this and in the following chapter. We write $\Gamma_P(f)$ for the set of 'pointwise' asymptotic values of $f(z)$ and

$$\Gamma_P(f, \theta_1 < \theta < \theta_2) = \bigcup_{\theta_1 < \theta < \theta_2} \Gamma(f, e^{i\theta})$$

for the subset relative to the interval (θ_1, θ_2) and, with this definition, we prove the following lemma.

LEMMA 1. *Suppose that $f(z)$ is meromorphic in $|z| < 1$ and regular and bounded in a simply connected domain D in $|z| < 1$ whose frontier consists of an arc $\Theta_1 \leqslant \theta \leqslant \Theta_2, z = e^{i\theta}$, and a Jordan arc in $|z| < 1$ joining the points $e^{i\Theta_1}$ and $e^{i\Theta_2}$. Then, for any pair θ_1, θ_2 such that $\Theta_1 < \theta_1 < \theta_2 < \Theta_2$,*

$$w_1 = f(e^{i\theta_1}) = \Gamma(f, e^{i\theta_1}), \quad w_2 = f(e^{i\theta_2}) = \Gamma(f, e^{i\theta_2})$$

and
$$w_1 \neq w_2,$$

the projection on the open straight line L between w_1 and w_2 of the set $\Gamma_P(f, \theta_1 < \theta < \theta_2)$ of values of $f(e^{i\theta})$ in the open interval $\theta_1 < \theta < \theta_2$ includes all points on L, so that the set $\Gamma_P(f, \theta_1 < \theta < \theta_2)$ is of positive linear measure.

Let D be mapped conformally onto the unit disc $|\xi| < 1$, the mapping function being $\xi(z)$ and its inverse $z(\xi)$. The arc

$\Theta_1 \leqslant \theta \leqslant \Theta_2$, $z = e^{i\theta}$, denoted by α, then transforms into an arc $\Psi_1 \leqslant \psi \leqslant \Psi_2$, $\xi = e^{i\psi}$, denoted by β. The points of the set $\Gamma_P(f)$ of radial limit values of $f(z)$ on α transform into points of $\Gamma_P(\varphi)$ relative to β, where $\varphi(\xi) = f(z(\xi))$, and so, by Theorem 2.3,

$$\Gamma_P(\varphi, \psi_1 \leqslant \psi \leqslant \psi_2) = \Gamma_P(f, \theta_1 \leqslant \theta \leqslant \theta_2),$$

where $e^{i\psi_1}$ and $e^{i\psi_2}$ are the transforms of $e^{i\theta_1}$ and $e^{i\theta_2}$ respectively. Further, since $\varphi(\xi)$ is bounded, almost all points of the arc β belong to the set $F(\varphi)$ of Fatou points of $\varphi(\xi)$, and almost all points of the arc α belong to $F(f)$, so that, as we have observed, the sets $\alpha \cap F(f)$ and $\beta \cap F(\varphi)$ are in one-to-one correspondence under the mapping.

Now we can find θ_1, θ_2 such that $\Theta_1 < \theta_1 < \theta_2 < \Theta_2$ and

$$w_1 = f(e^{i\theta_1}) \neq f(e^{i\theta_2}) = w_2,$$

where $e^{i\theta_1}$ and $e^{i\theta_2}$ both belong to $F(f)$ and w_1, w_2 to $\Gamma_P(f)$. For if not, $\varphi(\xi)$, which is bounded in $|\xi| < 1$ and has

$$\Gamma_P(\varphi, \Psi_1 \leqslant \psi \leqslant \Psi_2) = \Gamma_P(f, \Theta_1 \leqslant \theta \leqslant \Theta_2),$$

must be constant by Riesz's theorem (Theorem 2.5, Chapter 2), so that $f(z)$ is constant in D and hence also in $|z| < 1$.

We can find a circular arc λ in $|\xi| < 1$ joining the points $e^{i\psi_1}$, $e^{i\psi_2}$ on which $\varphi'(\xi) \neq 0$ and which is therefore mapped by $\varphi(\xi)$ onto a simple analytic arc Λ in the w-plane joining w_1, w_2. Let L be the rectilinear segment joining w_1, w_2 and let M be any straight line intersecting L perpendicularly. The number of points of intersection of Λ and M is finite and odd. There is therefore one point of intersection p such that the segment of M on one side of it has no other point of intersection with Λ and is mapped by the continuation of the ordinary element $e_\xi(\varphi, p)$ onto a path m in that portion of $|\xi| < 1$ cut off by λ and having the arc β in its boundary.

Now since $\varphi(\xi)$ is bounded, say $|\varphi(\xi)| < k$ for $|\xi| < 1$, and since $e_\xi(\varphi, p)$ is an internal element of $\xi(\varphi)$, there are two possibilities: either (i) this continuation contains external elements, in which case it contains a boundary element $e_\xi(\varphi, q)$, q being a point of M within the circle $|\varphi| < k$; or (ii) this continuation contains only internal elements, in which case it must be terminated by a

transcendent singularity ω on M within the circle $|\varphi| < k$. In case (i) the path pq is mapped by the continuation onto an asymptotic path in $|\xi| < 1$ having its end-point in the arc $\xi = e^{i\psi}, \psi_1 < \psi < \psi_2$, contained in β, q being a regular asymptotic value at this point. In case (ii) it follows from Theorem 6.2 that the path pw is mapped by the continuation onto an asymptotic path in $|\xi| < 1$ which, since $|\varphi(\xi)| < k$ for $|\xi| < 1$, terminates in a point in the arc $\xi = e^{i\psi}$, $\psi_1 < \psi < \psi_2$, contained in β and on which $\varphi(\xi) \to \omega$ as $|\xi| \to 1$. Thus, in case (i) $q \in \Gamma_P(f, \theta_1 < \theta < \theta_2)$ and in case (ii) $\omega \in \Gamma_P(f, \theta_1 < \theta < \theta_2)$ since $\varphi(\xi) = f(z(\xi))$. But M is any perpendicular to L, so that there is at least one point q or one point ω in $|w| < k$ on every M. This proves the lemma.

From Lemma 1 we have at once, by the remarks at the beginning of this paragraph,

THEOREM 6.3. *If $f(z)$ is meromorphic in $|z| < 1$ and $\mathscr{C}C(f)$ is not empty, then $\Gamma(f) = \Gamma_P(f)$ is of positive linear measure.*

In view of the form of Lemma 1 it is natural to ask whether the set $\Gamma_P(f)$ necessarily contains a continuum. The answer, as we shall prove later (Chapter 9, § 10), is in the negative, even in the case where every point of the circumference $|z| = 1$ belongs to the set $F(f)$ of Fatou points.

Evidently, the argument of Lemma 1 fails if $f(z)$ is not bounded in D, for the continuation along a line M may be carried to infinity for a set of positive measure on L, or the set $\Gamma_P(f)$ may be empty as is the case, for example, if $f(z)$ has more than one asymptotic value approached on spiral paths. We return to the subject of spiral paths in § 4 below.

Lemma 1 and Theorem 6.3 may thus be regarded as best possible.

3. Boundary theorems in the large

Our principal objective in this section is to prove the analogue in the large of Iversen's theorem (Theorem 1.6, Chapter 1) which states that, in the neighbourhood of an isolated essential singularity z_0 of a meromorphic function $f(z)$, an exceptional value $a \in \mathscr{C}R(f, z_0)$ is also an asymptotic value. Our immediate problem is to investigate the relations between the sets $C(f)$, $\mathscr{C}R(f)$ and

$\Gamma(f)$ for a function $f(z)$ meromorphic in $|z| < 1$, but not otherwise restricted. The formal analogy between our main result (Theorem 6.4) and Iversen's theorem is surprisingly close considering the great diversity of behaviour near the boundary $|z| = 1$ which is possible.

We require two further lemmas, which are of some interest in themselves, based upon the results of §§ 2 and 3 above.

LEMMA 2. *If $f(z)$ is meromorphic in $|z| < 1$, and if*

$$a \in \mathscr{C}\overline{\Gamma}(f) \cap C(f),$$

then a is an interior point of $R(f)$.

For a given positive number σ, we consider the set of domains $G(a, \sigma)$ in $|z| < 1$ in which $|f(z) - a| < \sigma$ or, if $a = \infty$, $1/|f(z)| < \sigma$. The union $\Delta_\sigma(a)$ of the domains $G(a, \sigma)$ is the open set on which the σ-neighbourhood $U(a, \sigma)$ of a is mapped by the inverse of f. The frontier of a domain $G(a, \sigma)$ consists only of level curves, which we shall call contours, on which $|f(z) - a| = \sigma$ (or $1/|f| = \sigma$ if $a = \infty$) in $|z| < 1$ and points of the circumference $|z| = 1$. Either of these elements may be absent; but if the frontier of $G(a, \sigma)$ does not meet the circumference $|z| = 1$ we say that G is *bounded*, otherwise it is *unbounded*. The justification for this terminology is that in the hyperbolic metric G is bounded or unbounded.

We assume that $\Delta_\sigma(a)$ has contours. The frontier of a bounded $G(a, \sigma)$ consists of a finite set of closed contours while the frontier of an unbounded G may contain either open or closed contours or both to any number, finite or infinite. An open contour is one which meets $|z| = 1$ and is thus not closed in $|z| < 1$.

By hypothesis we can find $\epsilon > 0$ such that $\overline{U}(a, \epsilon) \cap \Gamma(f)$ is empty.† Then for all $\sigma < \epsilon$ there is either (i) an infinity of bounded domains $G(a, \sigma)$ or (ii) at least one unbounded domain $G(a, \sigma)$. For if for some $\sigma < \epsilon$ neither of these conditions is realized there is an annulus $1 - \eta < |z| < 1$ in which $|f - a| \geqslant \sigma$ so that $a \in \mathscr{C}C(f)$. In case (i) each closed bounded $\overline{G}(a, \sigma)$ contains a zero of $f(z) - b$ for every b in $|b - a| \leqslant \sigma$, so that $\overline{U}(a, \sigma) \subset R(f)$.

We treat case (ii) in two stages. If the set of domains $G(a, \sigma)$ has only a finite number of closed contours and no open contour,

† We may assume that $a \neq \infty$.

there can be only one unbounded domain $G(a, \sigma)$ which, for a sufficiently small $\eta > 0$, contains an annulus $1 - \eta < |z| < 1$ in which $|f(z) - a| < \sigma$. It follows that $\Gamma(f) \subset U(a, \epsilon)$ and, by Theorem 6.3, that $\Gamma(f) = \Gamma(f) \cap U(a, \epsilon)$ is of positive linear measure. Since this is contrary to hypothesis, it follows that if the bounded domains $G(a, \sigma)$ are finite in number and if there is no unbounded domain $G(a, \sigma)$ having an open contour, there is an unbounded domain $G(a, \sigma)$ having an infinity of closed contours.

We now consider these two cases. First, if for a positive $\sigma < \epsilon$ there is a $G(a, \sigma)$ having an open contour, we can choose a point z on it and continue the corresponding inverse element $e_z(w', w)$, which is an internal element of $z(w)$, the inverse of $w = f(z)$, indefinitely round the circumference $|w - a| = \sigma$ in either direction. For, since $\Gamma(f) \cap \bar{U}(a, \epsilon)$ is empty by hypothesis, the continuation contains no boundary element and there can be no transcendent point of the internal branch of $z(w)$ on the circumference $|w - a| = \sigma$. It follows that for any b on $|b - a| = \sigma$ there is an infinity of zeros of $f(z) - b$ on the open contour in question, so that $b \in R(f)$. Secondly, if there is a domain $G(a, \sigma)$ having an infinity of closed contours, there is a zero of $f(z) - b$, $|b - a| = \sigma$, on each of them so that again $b \in R(f)$. Since $\sigma < \epsilon$, but is otherwise arbitrary, we conclude that every point of $U(a, \epsilon)$, except perhaps the point a, belongs to $R(f)$. We now show that $a \in R(f)$ also. Let a' be any point in $U(a, \tfrac{1}{2}\epsilon)$ and put $\sigma' = |a' - a| < \tfrac{1}{2}\epsilon$. Then $U(a', \tfrac{1}{2}\epsilon) \cap \Gamma(f)$ is empty, and we apply the foregoing argument to the domains $G(a', \sigma')$. Since $a' \in R(f) \subset C(f)$, case (i) is excluded and we have case (ii), so that $a \in R(f)$. We have thus proved that in case (ii) $U(a, \epsilon) \subset R(f)$ and this proves the lemma.

It is convenient at this point to introduce a new notation. We define the set $\Gamma_+(f) \subseteq \bar{\Gamma}(f)$ as follows: $a \in \Gamma_+(f)$ if, for all $\epsilon > 0$, $U(a, \epsilon) \cap \Gamma(f)$ is of positive linear measure. A point of $\Gamma_+(f)$ is not necessarily a point of $\Gamma(f)$.

LEMMA 3. *If $f(z)$ is meromorphic in $|z| < 1$ and if*

$$a \in \mathscr{C}\Gamma_+(f) \cap C(f),$$

then a is an interior point of $\bar{R}(f)$. If, further, $a \in \mathscr{C}\Gamma(f)$, then $a \in R(f)$.

Exactly as in the proof of Lemma 2, if $a \in C(f)$, then for all positive $\sigma < \epsilon$ there is either an infinity of bounded domains $G(a, \sigma)$ or an unbounded $G(a, \sigma)$ having either an open contour or an infinity of closed contours.

Since $\Gamma(f) \cap U(a, \epsilon)$ is of linear measure zero for some $\epsilon > 0$, the circumference $\gamma(\sigma_1)$ defined by $|w - a| = \sigma_1 < \epsilon$ for almost all σ_1 in $0 < \sigma_1 < \epsilon$ has no point of $\Gamma(f)$ on it. We shall assume that $\gamma(\sigma_1)$ has this property. If there is an infinity of closed contours of domains $G(a, \sigma_1)$, either bounded or unbounded, every value b on $\gamma(\sigma_1)$ belongs to $R(f)$. If there are only finitely many closed contours, there is at least one open contour of an unbounded $G(a, \sigma_1)$. Choose a point z on such a contour and let $e_z(w', w)$, where $w \in \gamma(\sigma_1)$, be the corresponding element of the inverse $z(w)$ of $f(z)$. We can continue $e_z(w', w)$ indefinitely round $\gamma(\sigma_1)$, the continuation containing only internal elements.† It follows again that every value $b \in \gamma(\sigma_1)$ belongs to $R(f)$. Since every point of $U(a, \epsilon)$ is arbitrarily near to a circumference $\gamma(\sigma_1)$, it follows that $U(a, \epsilon) \subseteq \bar{R}(f)$. It will be observed that this does not require the condition $a \in \mathscr{C}\Gamma(f)$ which is required only for the final step in the proof.

It now remains to prove that $a \in \mathscr{C}\Gamma(f)$ implies $a \in R(f)$. Suppose to the contrary that $a \in \mathscr{C}R(f)$. We can then find $\epsilon(a) > 0$ such that for all $\sigma_1 < \epsilon(a)$ all the zeros of $f(z) - a$ are contained in a finite set of bounded domains $G(a, \sigma_1)$. But since $a \in R'(f)$, the derived set of $R(f)$, we can find $b \in R(f)$ such that $|b - a| < \sigma_1$. Hence there is an unbounded domain $G_1(a, \sigma_1)$ which contains no zero of $f(z) - a$. Now, since $\Gamma(f) \cap U(a, \epsilon)$ is of linear measure zero and since $a \in \mathscr{C}\Gamma(f)$, we can find a diameter τ of the circle $\gamma(\sigma_1)$ on which there is no point of $\Gamma(f)$ and on which therefore every internal element of the inverse $z(w)$ of $f(z)$ at an end-point of τ on $\gamma(\sigma_1)$ can be continued through the point a to the antipodal point, the continuation again containing only internal elements. Choose a point z_1 on a contour of $G_1(a, \sigma_1)$ and continue the corresponding internal element $e_{z_1}(w, w_1)$, $|w_1 - a| = \sigma_1$, along the

† In fact a continuation of an element $e_z(w', w)$ along any path in $U(a, \epsilon)$ contains no boundary element since, by Lemma 1, if there were such an element, $U(a, \epsilon) \cap \Gamma(f)$ would be of positive linear measure because $f(z)$ would be bounded on an arc of $|z| = 1$ containing the corresponding regular point of $f(z)$.

circumference $\gamma(\sigma_1)$ to an end-point of τ which is mapped on a point $z(\tau)$ of the contour. The corresponding element $e_{z(\tau)}$ can be continued along τ through the point a to the antipodal point and in this way is mapped on a cross-cut of the domain $G_1(a, \sigma_1)$ on which there is a zero of $f(z) - a$. But $G_1(a, \sigma_1)$ contains no zero of $f(z) - a$, so that we have a contradiction, and we conclude that $a \in R(f)$. This completes the proof of the lemma.

COROLLARY. *If $C(f)$ is nowhere dense, then $C(f) \subset \Gamma_+(f)$.*

For $\bar{R}(f)$ has no interior points so that every $a \in C(f)$ belongs to $\Gamma_+(f)$. In particular, if $R(f)$ is empty, as in the case of a univalent function, $C(f) = \operatorname{Fr} C(f) \subset \Gamma_+(f)$ so that the accessible points of the boundary of any simply connected domain in the neighbourhood of any point of the boundary form a set of positive linear measure.

LEMMA 4. *If $f(z)$ is meromorphic in $|z| < 1$, then*

$$\operatorname{Fr} R(f) \cup \operatorname{Fr} C(f) = \overline{\mathscr{C}R(f)} \cap C(f).$$

First, $\quad \operatorname{Fr} R(f) \subset \overline{\mathscr{C}R(f)}$ and $\operatorname{Fr} R(f) \subset C(f)$

so that $\quad \operatorname{Fr} R(f) \subset \overline{\mathscr{C}R(f)} \cap C(f)$.

Also $\quad \operatorname{Fr} C(f) \subset \overline{\mathscr{C}C(f)} \cap C(f) \subset \overline{\mathscr{C}R(f)} \cap C(f)$.

Therefore $\quad \operatorname{Fr} R(f) \cup \operatorname{Fr} C(f) \subset \overline{\mathscr{C}R(f)} \cap C(f)$.

The opposite inclusion also holds; for if $\alpha \in \overline{\mathscr{C}R(f)} \cap C(f)$ and $\alpha \notin \operatorname{Fr} R(f) \cup \operatorname{Fr} C(f)$, then $\alpha \in \operatorname{Int} C(f) \subset \bar{R}(f)$, by Theorem 4.2. Since also $\alpha \notin \operatorname{Fr} R(f)$ we must have $\alpha \in \operatorname{Int} R(f)$ which is not in $\overline{\mathscr{C}R(f)}$. (Here $\operatorname{Int} M$ denotes the interior of the set M.)

We are now in a position to prove our main theorem in the large, namely

THEOREM 6.4. *If $f(z)$ is meromorphic in $|z| < 1$, then the following relations are satisfied:*

(i) *if $\Gamma(f)$ is unrestricted,*

$$\operatorname{Fr} R(f) \cup \operatorname{Fr} C(f) = \overline{\mathscr{C}R(f)} \cap C(f) \subseteq \bar{\Gamma}(f); \qquad (6.3.1)$$

(ii) *if $\Gamma(f)$ is of linear measure zero,*

$$\mathscr{C}R(f) \subseteq \Gamma(f). \qquad (6.3.2)$$

To prove (i) we use Lemma 2. By that lemma

$$\mathscr{C}\bar{\Gamma}(f) \subseteq \mathscr{C}C(f) \cup \operatorname{Int} R(f), \qquad (6.3.3)$$

where $\operatorname{Int} R(f)$ denotes the interior of $R(f)$, so that, by taking complements, we have

$$\overline{\mathscr{C}R(f)} \cap C(f) \subseteq \bar{\Gamma}(f). \qquad (6.3.4)$$

The complete relation (6.3.1) now follows from Lemma 4.

To prove (ii) we use Lemma 3. By that lemma

$$\mathscr{C}\Gamma(f) \cap \mathscr{C}\Gamma_+(f) \subseteq \mathscr{C}C(f) \cup R(f)$$

and hence $\qquad \mathscr{C}R(f) \subseteq \mathscr{C}C(f) \cup \Gamma(f) \cup \Gamma_+(f). \qquad (6.3.5)$

But since, by hypothesis, $\Gamma(f)$ is of linear measure zero, $\Gamma_+(f)$ is empty, and $\mathscr{C}C(f)$ is also empty by Theorem 6.3.

It can be shown by examples that Theorem 6.4 (i) *is best possible* in the sense that $\bar{\Gamma}(f)$ cannot be replaced in (6.3.1) by $\Gamma(f)$, so that the formulation of Theorem 1.6, Chapter 1, is not generally applicable in the large. The simplest example is that of a univalent function $f(z)$ which maps the unit disc $|z| < 1$ onto a domain which has just one prime end not of the first kind,† which we may take to correspond to the point $z = 1$. Let the domain be the square D cut along the vertical lines of length $1 - (1/n)$ erected at the points $(1 - (1/n), 0)$, $n \geqslant 2$, as indicated in figure 6.1, converging to the line $[1, 1 + i]$, which is the impression of the prime end of D, in this case of the second kind, having the point $1 + i$ as its accessible point. The point $w_0 = 1 + \frac{1}{2}i$ is an inaccessible point on the frontier of D and therefore belongs to $C(f)$ but not to $\Gamma(f)$. But w_0 is the limit of the points

$$\left(1 - \frac{1}{n}, \frac{i}{2}\right),$$

each of which is accessible from both sides of the cut erected on the point $(1 - (1/n), 0)$, and is therefore a point of $\bar{\Gamma}(f)$. Since $f(z)$ is univalent, $R(f)$ is empty, and consequently $\underline{\mathscr{C}R(f)} = \overline{\mathscr{C}R(f)}$ is total. Therefore $w_0 \in \mathscr{C}R(f) \cap C(f)$, while $w_0 \in \bar{\Gamma}(f) \cap \mathscr{C}\Gamma(f)$.

† The classification of prime ends is given in Chapter 9, §8.

The above example is of a very special kind in that $R(f)$ is empty. It is, however, in principle a simple matter, although the details are rather long, to construct an automorphic function $f(z)$ having the same property, that there exists a value

$$w_0 \in \mathscr{C}R(f) \cap C(f) \cap \overline{\Gamma}(f) \cap \mathscr{C}\Gamma(f),$$

but for which $R(f)$ is everywhere dense and $\overline{\mathscr{C}R(f)}$ is nowhere dense.†

Fig. 6.1.

We proceed to deduce some of the immediate consequences of Theorem 6.4.

COROLLARY 1. *If $f(z)$ is meromorphic in $|z| < 1$, a necessary and sufficient condition for*

$$\mathscr{C}R(f) \subseteq \overline{\Gamma}(f) \tag{6.3.6}$$

is that $R(f)$ be everywhere dense; and if a value $a \in \mathscr{C}R(f)$ is not in $\overline{\Gamma}(f)$, then $a \in \mathscr{C}C(f) \subseteq \operatorname{Int}\mathscr{C}R(f)$.

The condition is necessary, since, by (6.3.1),

$$\mathscr{C}R(f) \cap C(f) \subseteq \overline{\mathscr{C}R(f)} \cap C(f) \subseteq \overline{\Gamma}(f),$$

and from this follows

$$\mathscr{C}R(f) \cap \mathscr{C}\overline{\Gamma}(f) \subseteq \mathscr{C}C(f).$$

† Collingwood and Cartwright [1], pp. 113–116.

The condition is sufficient, since it implies that

$$\mathscr{C}R(f) \subseteq \mathrm{Fr}\, R(f) \subseteq \bar{\Gamma}(f).$$

COROLLARY 2. *If $f(z)$ is meromorphic in $|z| < 1$ and $T(r,f)$ is unbounded, then $\mathscr{C}R(f) \subseteq \bar{\Gamma}(f)$.*

For, by Theorem 4.1, $C(f)$ is total and therefore, by Theorem 4.2, $R(f)$ is everywhere dense.

Other corollaries follow if further conditions are imposed on the set $\Gamma(f)$. Thus we have, for example, recalling Theorem 2.20, and the definition of $(\Phi(f))$ there given,

COROLLARY 3. *If $\mathscr{C}R(f)$ is infinite then $\Gamma(f)$ is infinite, and if $\mathscr{C}R(f)$ is not an isolated set then $\Gamma(f)$ is not an isolated set. In either case $\Gamma(f) = \Gamma_P(f)$ and $\Phi(f)$ is empty.*

Corollary 3 is illustrated by the modular function $\mu(z)$, regular in $|z| < 1$, for which $\mathscr{C}R(\mu)$ and $\Gamma(\mu) = \Gamma_P(\mu)$ both consist of the points 0, 1, ∞ and by $Q(z) = \log \mu(z)$ for which $\mathscr{C}R(Q)$ and $\Gamma(Q)$ both consist of ∞, $\pm 2n\pi i$ ($n = 0, 1, 2, \ldots$).†

For further results following from Theorem 6.4 we refer the reader to Collingwood and Cartwright [1].

4. Cluster sets on spiral paths

We shall confine ourselves for convenience to the case of spiral paths in the open disc $|z| < 1$ converging to the circumference K: $|z| = 1$, but the methods we use and the results we shall prove are equally valid for more general classes of paths which converge to the whole circumference K from inside $|z| < 1$. The theory of cluster sets on such curves clearly belongs to the theory in the large.

We have already seen (Theorem 4.7, Chapter 4), that, given an *arbitrary* function $f(z)$, defined and one-valued in $|z| < 1$, there is an isolated set \mathscr{G} of points

$$\{z_n\}, \quad |z_n| < |z_{n+1}|, \quad \lim_{n\to\infty} |z_n| = 1,$$

such that $C_{\mathscr{G}}(f) = C(f)$, from which it follows that, given any spiral σ, which we may assume to be monotonic and passing through the points $\{z_n\}$ in succession (and therefore converging

† See J. E. Littlewood [2], p. 185.

to K), $C_\sigma(f) = C(f)$. This establishes quite trivially the existence of spirals of maximum indetermination for any function. We shall show, however, that for certain classes of meromorphic functions, *every* spiral is a path of maximum indetermination, while rather weaker but still significant results can be established for certain other classes.

We prove the following theorem.

THEOREM 6.5. *If $f(z)$ is meromorphic in $|z| < |$ and if $\mathscr{C}R(f)$ contains three values, then, for any spiral path σ in $|z| < 1$ converging to K,*

$$C_\sigma(f) \supseteq C(f) \cap \overline{\mathscr{C}R(f)}. \qquad (6.4.1)$$

Because σ is a spiral converging to K and $C_\sigma(f)$ is closed, we have $\overline{\Gamma}_P(f) \subset C_\sigma(f)$. But, by Theorems 6.4 (i) and 2.20,

$$\overline{\mathscr{C}R(f)} \cap C(f) \subset \overline{\Gamma}(f) = \overline{\Gamma}_P(f)$$

since $\mathscr{C}R(f)$ contains three values so that $\Gamma_A(f)$ is empty. Here $\Gamma_A(f)$ denotes the set of 'arcwise' asymptotic values achieved on paths whose ends are not points, but arcs of K, so that $\Gamma_A(f) \subset \Phi(f)$. Plainly $\Gamma_P(f) \cup \Gamma_A(f) = \Gamma(f)$.

We note that if $\mathscr{C}R(f)$ contains three values, C_σ is always a continuum and thus σ cannot be an asymptotic path.

COROLLARY 1. *If $f(z)$ is meromorphic in $|z| < 1$ and if its Dirichlet integral*

$$\iint_{|z|<1} |f'(z)|^2\, dx\, dy$$

is finite, then on any spiral σ converging to K,

$$C_\sigma(f) = C(f). \qquad (6.4.2)$$

By hypothesis $\mathscr{C}R(f)$ must be dense. Otherwise there is a domain \mathscr{D} on the w-sphere contained in $R(f)$, and therefore covered infinitely often by the Riemann surface generated by $f(z)$, the area of which, defined by the Dirichlet integral, is infinite. Hence $\overline{\mathscr{C}R(f)}$ is total and since $C_\sigma(f) \subseteq C(f)$, (6.4.2) follows from (6.4.1).

COROLLARY 2. *If $f(z)$, meromorphic in $|z| < 1$, is such that $C(f) = \operatorname{Fr} C(f)$, then* $$C_\sigma(f) = C(f)$$ *for any spiral σ in $|z| < 1$ converging to K.*

For $\mathscr{C}R(f)$ must be dense, and the corollary is immediate.

In particular, (6.4.2) holds for every function of finite valency, as well as for functions for which

$$\iint_{|z|<1} \frac{|f'|^2}{[1+|f|^2]^2}\,dx\,dy$$

is finite.

COROLLARY 3. *If $f(z)$ is such that $C(f)$ is subtotal, then for any spiral σ in $|z| < 1$ converging to K,*

$$C_\sigma(f) \supseteq \text{Fr}\,C(f). \tag{6.4.3}$$

For, $\text{Fr}\,C(f) \subseteq \overline{\mathscr{C}R(f)}$, and $\text{Fr}\,C(f)$ is not empty since $\mathscr{C}C(f)$ is not empty.

This corollary is best possible as is shown by the example of the Blaschke product

$$B(z) = \prod_{n=1}^{\infty} \frac{z-t_n}{1-t_n z},$$

where

$$t_n = \frac{n!-1}{n!+1} \quad (n = 1, 2, \ldots).$$

It can be shown† that, if we consider the points

$$\tau_n = \frac{(n+1)!-\rho}{(n+1)!+\rho} \quad (1 < \rho < 2, n = 1, 2, \ldots)$$

lying between successive zeros t_n of the function $B(z)$, then

$$\lim_{n\to\infty} |B(\tau_n)| = \frac{\rho-1}{\rho+1}.$$

A similar argument applied to the sequence of points

$$\tau_n' = \frac{(n+1)!-(n+1)}{(n+1)!+(n+1)},$$

which also separates successive zeros of $B(z)$, shows that

$$\lim_{n\to\infty} B(\tau_n') = 1.$$

The point $z = 1$ is the only singularity of $B(z)$ on K so that $|B(z)| = 1$ at every other point $z \in K$. It follows that for any $\alpha < 1$

† Seidel [1], p. 14.

the components of the open set in which $|B(z)| < \alpha$ are infinite in number and are all contained, together with the level curves which are their boundaries, in the open disc $|z| < 1$ and are symmetric with respect to the real axis. We can therefore trace a spiral σ converging to K whose only intersections with the positive real axis are points of the sequence $\{\tau'_n\}$ and on which $|B(z)| \to 1$ as $z \to 1$. Since the components just referred to are infinite in number, $C(B)$ is the closed unit disc $|w| \leqslant 1$, so that it follows from Corollary 3 that $C_\sigma(B) = \text{Fr}\, C(B)$, which proves our assertion.

The theory of spiral paths has made considerable advances and has become too extensive for anything like a full treatment here. For further results bearing on this topic, the reader is referred to Bagemihl and Seidel [1 a] and [2], Seidel [3], Collingwood [13] and McMillan [1].

CHAPTER 7

BOUNDARY THEORY IN THE SMALL

1. Introduction

The methods of Chapter 6 are applicable not only in the large but also locally, or, as it is sometimes expressed, in the small. In this chapter we develop the local theory by these methods.

We begin by giving the definitions and proving the lemmas and collateral results analogous to those of the theory in the large which are required to prove the principal result of the present chapter, namely Theorem 7.2. We need to be able to describe the behaviour of $f(z)$, which will be assumed to be meromorphic in $|z| < 1$, near a given point $z = e^{i\theta}$. For this purpose we adopt the following additional notations and definitions.

We shall say that
$$a \in \Gamma(f, \theta_1 < \theta < \theta_2),$$
or
$$a \in \Gamma(f, \theta_1 \leqslant \theta \leqslant \theta_2),$$

if there is an asymptotic path on which $f(z)$ tends to a and whose *end*, that is to say, the intersection of its closure with
$$K = \{z : |z| = 1\},$$

is contained in the open arc $z = e^{i\theta}$, $\theta_1 < \theta < \theta_2$, or the closed arc $z = e^{i\theta}$, $\theta_1 \leqslant \theta \leqslant \theta_2$, as the case may be. The subsets
$$\Gamma_P(f, \theta_1 < \theta < \theta_2) \quad \text{and} \quad \Gamma_P(f, \theta_1 \leqslant \theta \leqslant \theta_2)$$

were defined in Chapter 6. For brevity we write
$$\Gamma(f, \theta - \eta < \theta' < \theta + \eta) = \Gamma(f, |\theta - \theta'| < \eta)$$

and, in particular,
$$\Gamma(f, -\eta < \theta < \eta) = \Gamma(f, |\theta| < \eta);$$

similarly for Γ_P and Γ_A, the subsets of *pointwise* and *arcwise* asymptotic values.

The intersection of the sets $\Gamma(f, |\theta' - \theta| < \eta)$ for all positive η, which we denote by

$$\chi(f, e^{i\theta}) = \bigcap_{\eta>0} \Gamma(f, |\theta' - \theta| < \eta) \qquad (7.1.1)$$

plays a rôle in the theory in the small similar to that of $\Gamma(f)$ in the theory in the large. We also write

$$\chi_P(f, e^{i\theta}) = \bigcap_{\eta>0} \Gamma_P(f, |\theta' - \theta| < \eta)$$

and

$$\chi_A(f, e^{i\theta}) = \bigcap_{\eta>0} \Gamma_A(f, |\theta' - \theta| < \eta),$$

so that

$$\chi(f, e^{i\theta}) = \chi_P(f, e^{i\theta}) \cup \chi_A(f, e^{i\theta}).$$

The sets $\chi_P(f, e^{i\theta})$ and $\chi_A(f, e^{i\theta})$, either or both of which may be empty, are the local analogues, for the purposes of the theory to be developed, of the global sets $\Gamma_P(f)$ and $\Gamma_A(f)$. Evidently $\Gamma_P(f, e^{i\theta}) \subseteq \chi_P(f, e^{i\theta})$.

As the analogue in the small of $\overline{\Gamma}(f)$ we define the set

$$\chi^*(f, e^{i\theta}) = \bigcap_{\eta>0} \overline{\Gamma}(f, |\theta' - \theta| < \eta). \qquad (7.1.2)$$

Evidently

$$\chi'(f, e^{i\theta}) \subseteq \bigcap_{\eta>0} \Gamma'(f, |\theta' - \theta| < \eta),$$

where χ' and Γ' are the derived sets of χ and Γ, so that

$$\chi(f, e^{i\theta}) \subseteq \overline{\chi}(f, e^{i\theta}) \subseteq \chi^*(f, e^{i\theta}). \qquad (7.1.3)$$

The set $\chi(f, e^{i\theta})$ may be empty even though $\Gamma(f, |\theta' - \theta| < \eta)$ is not empty for any $\eta > 0$; but $\chi^*(f, e^{i\theta})$ is not empty unless $\Gamma(f, |\theta' - \theta| < \eta)$ is empty for some $\eta > 0$. Plainly

$$\chi(f, e^{i\theta}) \subseteq \chi^*(f, e^{i\theta}) \subseteq C(f, e^{i\theta}). \qquad (7.1.4)$$

Now

$$\Gamma(f, |\theta' - \theta| < \eta) = \Gamma_P(f, |\theta' - \theta| < \eta) \cup \Gamma_A(f, |\theta' - \theta| < \eta)$$
$$= \Gamma_P(f, e^{i\theta}) \cup \Gamma_P(f, 0 < |\theta' - \theta| < \eta)$$
$$\cup \Gamma_A(f, |\theta' - \theta| < \eta); \qquad (7.1.5)$$

moreover

$$\Gamma_A(f, |\theta' - \theta| < \eta) \subseteq \Gamma_A(f, 0 < |\theta' - \theta| < \eta) \cup \Phi(f, e^{i\theta}), \qquad (7.1.6)$$

where $\Phi(f, e^{i\theta})$ is the local analogue of the set $\Phi(f)$ defined in

Chapter 2, p. 42. For if there is an arcwise asymptotic path on which $f(z) \to a$ and whose end contains $e^{i\theta}$, then $a \in \Phi(f, e^{i\theta})$. Combining (7.1.5) and (7.1.6) we have

$$\Gamma(f, |\theta' - \theta| < \eta)$$
$$\subseteq \Gamma(f, 0 < |\theta' - \theta| < \eta) \cup \Gamma_P(f, e^{i\theta}) \cup \Phi(f, e^{i\theta}), \quad (7.1.7)$$

and so†

$$\chi(f, e^{i\theta}) = \bigcap_{\eta > 0} \Gamma(f, |\theta' - \theta| < \eta)$$
$$\subseteq \Psi(f, e^{i\theta}) \cup \Gamma_P(f, e^{i\theta}) \cup \Phi(f, e^{i\theta}) \quad (7.1.8)$$

and

$$\chi^*(f, e^{i\theta}) = \bigcap_{\eta > 0} \overline{\Gamma}(f, |\theta' - \theta| < \eta)$$
$$\subseteq \Psi^*(f, e^{i\theta}) \cup \overline{\Gamma}_P(f, e^{i\theta}) \cup \Phi(f, e^{i\theta}). \quad (7.1.9)$$

By analogy with the definition of the set $\Gamma_+(f)$ we say that $a \in \chi_*(f, e^{i\theta})$ if for all $\eta > 0$ and $\epsilon > 0$ the set

$$U(a, \epsilon) \cap \Gamma(f, |\theta' - \theta| < \eta)$$

is of positive linear measure. This does not imply that

$$U(a, \epsilon) \cap \chi(f, e^{i\theta})$$

is of positive linear measure, which is the reason for not using the subscript $+$ as in the case of $\Gamma_+(f)$. But $a \in \mathscr{C}\chi_*(f, e^{i\theta})$ does imply that for some $\epsilon_0 > 0$, $\eta_0 > 0$, $U(a, \epsilon_0) \cap \Gamma(f, |\theta' - \theta| < \eta_0)$ is of linear measure zero and therefore $U(a, \epsilon_0) \cap \chi(f, e^{i\theta})$ is of linear measure zero.

In § 4 we extend the classification of singularities of a function on the circumference of the disc and prove a number of theorems, some of them quite recent, on the relative distribution of these classes of points in relation to the set of Fatou points and the corresponding set of Fatou values of the function. In § 5 we prove a recent theorem of Meier which is a very interesting analogue of Plessner's theorem, in which the excepted set instead of being of measure zero is of first category.

2. The main theorem in the small

We have now set up the necessary apparatus and proceed to prove the preliminary lemmas. These are very similar both in form and method of proof to those proved in Chapter 6.

† For definitions of Ψ and Ψ^* see the index of symbols.

LEMMA 1. *If $f(z)$ is meromorphic in $|z| < 1$ and if*

$$a \in \mathscr{C}\chi^*(f, e^{i\theta}) \cap C(f, e^{i\theta}),$$

then either

(i) *a is an interior point of $R(f, e^{i\theta})$; or*

(ii) *$a \in \Phi(f, e^{i\theta})$ and, for some $\epsilon > 0$,*

$$U(a, \epsilon) - \{a\} \subset R(f, e^{i\theta}).$$

We put $\theta = 0$ and denote by $D = D(1, \eta)$ the domain

$$|z - 1| < 2 \sin \tfrac{1}{2}\eta, \quad |z| < 1,$$

cut off from the unit disc by a circle of centre $z = 1$ through the points $z = e^{\pm i\eta}$, and by $\Sigma = \Sigma(a, \sigma, \eta)$ the set of domains $G(a, \sigma) \cap D$.

By hypothesis we can find $\sigma > 0$ and $\eta > 0$ such that $U(a, \sigma) \cap \Gamma(f, |\theta| < \eta)$ is empty. We now have to consider the various possibilities arising from this and the condition $a \in C(f, 1)$.

We note first that in every case $z = 1$ is a limit point of contours of Σ. For if not then either $|f(z) - a| \geqslant \sigma$ (or $1/|f(z)| \geqslant \sigma$ if $a = \infty$) in D for some positive σ and η in which case $a \in \mathscr{C}C(f, 1)$, or $\Sigma = D$ in which case $|f(z) - a| < \sigma$ in D and hence

$$U(a, \sigma) \cap \Gamma(f, |\theta| < \eta)$$

is of positive measure by Lemma 1 on p. 120.

This leaves us with the following two alternatives:

(i) For all $\sigma < \epsilon_0 \leqslant \epsilon$ the frontier of Σ contains either an infinity of closed contours of domains $G(a, \sigma)$ or an open contour having an end in the arc $z = e^{i\theta}$, $|\theta| < \eta$. It then follows from the argument used in the proof of the second part of Lemma 2 of Chapter 6 that, since there is no point of $\Gamma(f, |\theta| < \eta)$ on the circumference $\gamma(\sigma)$ defined by $|w - a| = \sigma$, there is an infinity of zeros of $f(z) - b$ in D for every $b \in \gamma(\sigma)$, so that $b \in R(f, 1)$ and $a \in R(f, 1)$.

(ii) The frontier of Σ does not contain either an infinity of closed contours of domains $G(a, \sigma)$ or an open contour having an end in the arc $z = e^{i\theta}$, $|\theta| < \eta$, but contains an infinity of open contours in D having no ends in this arc and therefore having their end-points on the arc $|z - 1| = 2 \sin \tfrac{1}{2}\eta$, $|z| < 1$. It is easily

seen that since $f(z)$ is meromorphic, only the points $e^{\pm i\eta}$ can be limit points of these end-points. For suppose, on the contrary, that a point z_0 with $|z_0| < 1$, $|z_0 - 1| = 2\sin\frac{1}{2}\eta$, is a limit point of contours of Σ which also have $z = 1$ as a limit point. Then, by continuity, $|f(z_0) - a| = \sigma$, and we can find positive numbers $\delta < \sin\frac{1}{2}\eta$ and $\beta < \sigma$ such that $0 < |f(z) - f(z_0)| < \beta$ for all $0 < |z - z_0| < \delta$. Now let $\beta_0 = \min |f(z) - f(z_0)|$ on $|z - z_0| = \delta$. The domain $G_0(a, \beta_0)$ containing z_0 therefore lies in $|z - z_0| < \delta$ and by our choice of δ we can ensure that its frontier lies wholly inside $|z| < 1$. But this frontier $\mathrm{FR}\, G_0(a, \beta_0)$ intersects infinitely many of the arcs of the frontier $\mathrm{FR}\,\Sigma$ of Σ. Denote these points of intersection by $z_1', z_2', \ldots, z_n', \ldots$, and let w_1 and w_2 be the two points of intersection of the circle $|w - f(z_0)| = \beta_0$ with the circle $|w - a| = \sigma$. Then, since $|f(z) - f(z_0)| = \beta_0$ on $\mathrm{FR}\, G_0(a, \beta_0)$, we have $f(z_n') = w_1$ or w_2 for all values of n, so that there are infinitely many zeros either of $f(z) - w_1$ or of $f(z) - w_2$ on $\mathrm{FR}\, G_0(a, \beta_0)$ and consequently a limit point of one of these sets of zeros in $|z| < 1$. But this is impossible since $f(z)$ is meromorphic, and thus our assertion is proved. The same clearly holds for all smaller values of η, and it follows that such a set of contours converges to at least one of the arcs $z = e^{i\theta}$, $-\eta \leqslant \theta \leqslant 0$ or $0 \leqslant \theta \leqslant \eta$. If this condition is satisfied for any arbitrarily small $\sigma > 0$, then $a \in \Phi(f, 1)$; and so if $a \in \mathscr{C}\Phi(f, 1)$, this condition is not satisfied for $\sigma < \epsilon_0$, say. Suppose now that $a \in C(f, 1) \cap \mathscr{C}\Phi(f, 1)$. Then we can find ϵ_0 such that Σ satisfies condition (i) for all $\sigma < \epsilon_0$, and it follows by the argument above that $\gamma(\sigma) \subseteq R(f, 1)$ for all $0 < \sigma < \epsilon_0$. We now show that also $a \in R(f, 1)$. Choose a sequence $\sigma_1 > \sigma_2 > \ldots > \sigma_n > \ldots$, $\lim \sigma_n = 0$, $\sigma_1 < \epsilon_0$, and let b_1 be a point of $\gamma(\sigma_1)$. Since $b_1 \in R(f, 1)$ we can find a b_1-point z_1 of $f(z)$, i.e. a zero of $f(z) - b_1$, in $D(1, \frac{1}{2}\eta)$. The corresponding element $e_{z_1}(w, b_1)$ of the inverse function $z(w)$ can be continued along the radius of $\gamma(\sigma_1)$ to the centre a and the continuation maps this radius onto a curve k_1 in $|z| < 1$, one of whose end-points is the point z_1 while the other is an a-point. Similarly, let b_2 be a point of $\gamma(\sigma_2)$ not on the radius to b_1 and let z_2 be a b_2-point of $f(z)$ in $D(1, \frac{1}{2}\eta)$; we continue the corresponding element along the radius to a. The continuation maps this radius of $\gamma(\sigma_2)$ onto a curve k_2 joining z_2 to an a-point of $f(z)$. Repeat the process for b_3 on $\gamma(\sigma_3)$ but not on either of the radii to

b_1 and b_2 and a b_3-point z_3 in $D(1, \frac{1}{4}\eta)$, and generally for b_n on $\gamma(\sigma_n)$ and z_n in $D(1, \eta/(n+1))$. In this way we obtain a sequence of curves k_n for which $z = 1$ is a limit point and on which

$$|f(z) - a| < \sigma_n \quad (|f(z)| < 1/\sigma_n \quad \text{if} \quad a = \infty).$$

Now suppose that $a \in \mathscr{C}R(f, 1)$ and that η has been so chosen that $D(1, \eta)$ contains no a-point of $f(z)$. Then all the curves k_n cross the circumference $|z - 1| = 2 \sin \frac{1}{2}\eta$. Further, they have no limit point in $D(1, \eta)$ since at such a limit point z_0 we should have $f(z_0) = a$, contrary to hypothesis. Therefore the sequence of distinct curves k_n contains a subsequence which converges to at least one of the arcs $z = e^{i\theta}$, $-\eta \leqslant \theta \leqslant 0$, $0 \leqslant \theta \leqslant \eta$, so that $a \in \Phi(f, 1)$, which is a contradiction. Therefore, under the hypothesis of the lemma, $a \in C(f, 1) \cap \mathscr{C}\Phi(f, 1)$ implies that $U(a, \epsilon_0) \subseteq R(f, 1)$.

Finally, $a \in \Phi(f, 1)$ implies $U(a, \epsilon) - \{a\} \subseteq R(f, 1)$ for sufficiently small ϵ, since by Theorem 2.20 $\mathscr{C}R(f, 1)$ then contains at most two values. This completes the proof of the lemma.

As the analogue of Lemma 3 of Chapter 6 we prove

LEMMA 2. *If $f(z)$ is meromorphic in $|z| < 1$ and if*

$$a \in \mathscr{C}\chi_*(f, e^{i\theta}) \cap \mathscr{C}\chi(f, e^{i\theta}) \cap C(f, e^{i\theta}), \qquad (7.2.1)$$

then either

(i) *a is an interior point of $\bar{R}(f, e^{i\theta})$ and $a \in R(f, e^{i\theta})$; or*

(ii) *$a \in \Phi(f, e^{i\theta})$ and, for some $\epsilon > 0$, $U(a, \epsilon) - \{a\} \subset R(f, e^{i\theta})$.*

We set $\theta = 0$ and define D and Σ as in the proof of Lemma 1. Then, just as in that proof, if $a \in C(f, 1) \cap \mathscr{C}\Phi(f, 1)$ we can find ϵ_0 such that for all $\sigma < \epsilon_0$ the frontier of Σ contains either an infinity of closed contours or an open contour having an end in the arc $z = e^{i\theta}$, $|\theta| < \eta$. Now for almost all σ in $0 < \sigma < \epsilon_0$ there is no point of $\Gamma(f, |\theta| < \eta)$ on $\gamma(\sigma)$. Values of σ having this property we denote by σ_1. It follows that $\gamma(\sigma_1) \subseteq R(f, 1)$ and hence that $U(a, \epsilon_0) \subseteq \bar{R}(f, 1)$. To prove that $a \in R(f, 1)$, we now choose a sequence $\sigma_{11} > \sigma_{12} > \ldots > \sigma_{1n} > \ldots$, $\lim \sigma_{1n} = 0$, of numbers belonging to the set of σ_1 and, as we may do since $\Gamma(f, |\theta| < \eta)$ is of measure zero, we choose a point b_{1n} on each $\gamma(\sigma_{1n})$ such that there is no point of $\Gamma(f, |\theta| < \eta)$ and no other point $b_{1\nu}$, $\nu \neq n$, on the diameter of $\gamma(\sigma_{1n})$ through b_{1n}. We now

repeat for the points b_{1n} the argument of Lemma 1 (ii) which applies without modification.

We have only to note that in the penultimate paragraph of the proof the assumption $a \in \mathscr{C}R(f, 1)$ again implies that all the curves k_n cross the circumference $|z-1| = 2 \sin \frac{1}{2}\eta$ since, by the condition $a \in \mathscr{C}\chi(f, e^{i\theta})$, none of them can meet the arc $e^{i\theta}$, $|\theta| < \eta$, for sufficiently small η.

This proves the lemma.

The analogue in the small of Theorem 6.3 is the following.

THEOREM 7.1. *If $f(z)$ is meromorphic in $|z| < 1$ and if for some θ the set $\mathscr{C}C(f, e^{i\theta})$ is not empty, then the set*

$$\Gamma(f, |\theta'-\theta| < \eta) = \Gamma_P(f, |\theta'-\theta| < \eta) \qquad (7.2.2)$$

for sufficiently small $\eta > 0$ and is of positive linear measure, and thus the set $\chi_(f, e^{i\theta})$ is not empty.*

Put $\theta = 0$ and choose $a \in \mathscr{C}C(f, 1)$. Then we can find $\eta_0 > 0$ such that the function $1/(f(z) - a)$ is regular and bounded in $D(1, \eta_0)$. The theorem then follows immediately from Lemma 2.

A similar argument to that by which we proved Lemma 4 of Chapter 6 now gives us its local analogue:

LEMMA 3. *If $f(z)$ is meromorphic in $|z| < 1$, then for any θ,*

$$\mathrm{Fr}\,R(f, e^{i\theta}) \cup \mathrm{Fr}\,C(f, e^{i\theta}) = \overline{\mathscr{C}R(f, e^{i\theta})} \cap C(f, e^{i\theta}). \qquad (7.2.3)$$

We can now prove our main theorem in the small, namely

THEOREM 7.2.† *If $f(z)$ is meromorphic in $|z| < 1$, then for any θ, $0 \leqslant \theta \leqslant 2\pi$, the following relations are satisfied:*

(i) *if $\chi(f, e^{i\theta})$ is unrestricted,*

$$\mathrm{Fr}\,R(f, e^{i\theta}) \cup \mathrm{Fr}\,C(f, e^{i\theta}) = \overline{\mathscr{C}R(f, e^{i\theta})} \cap C(f, e^{i\theta})$$
$$\subseteq \chi^*(f, e^{i\theta}) \cup \Phi(f, e^{i\theta}); \qquad (7.2.4)$$

(ii) *if $\chi_*(f, e^{i\theta})$ is empty, i.e. if $\Gamma(f, |\theta'-\theta| < \eta)$ is of linear measure zero for some $\eta > 0$, then*

$$\mathscr{C}R(f, e^{i\theta}) \subseteq \chi(f, e^{i\theta}) \cup \Phi(f, e^{i\theta}), \qquad (7.2.5)$$

† A theorem closely related to Theorem 7.2 has recently been proved by McMillan [3]; see also Ryan [1, 2].

and if, further, $\mathscr{C}R(f, e^{i\theta})$ contains more than two values, then $\Phi(f, e^{i\theta})$ and $\chi_A(f, e^{i\theta})$ are both empty and

$$\mathscr{C}R(f, e^{i\theta}) \subseteq \chi(f, e^{i\theta}) = \chi_P(f, e^{i\theta}). \qquad (7.2.6)$$

To prove (i) we use Lemma 1. By that lemma,

$$\mathscr{C}\chi^*(f, e^{i\theta}) \subseteq \mathscr{C}C(f, e^{i\theta}) \cup \operatorname{Int} R(f, e^{i\theta}) \cup \Phi(f, e^{i\theta}),$$

and hence

$$\overline{\mathscr{C}R(f, e^{i\theta})} \cap C(f, e^{i\theta}) \subseteq \chi^*(f, e^{i\theta}) \cup \Phi(f, e^{i\theta}).$$

The complete relation (7.2.4) now follows from Lemma 3.

To prove (ii) we use Lemma 2. This gives

$$\mathscr{C}\chi(f, e^{i\theta}) \cap \mathscr{C}\chi_*(f, e^{i\theta}) \subseteq \mathscr{C}C(f, e^{i\theta}) \cup R(f, e^{i\theta}) \cup \Phi(f, e^{i\theta})$$

and hence

$$\mathscr{C}R(f, e^{i\theta}) \subseteq \mathscr{C}C(f, e^{i\theta}) \cup \chi(f, e^{i\theta}) \cup \chi_*(f, e^{i\theta}) \cup \Phi(f, e^{i\theta}). \quad (7.2.7)$$

Now, by Theorem 7.1, if $\chi_*(f, e^{i\theta})$ is empty, $\mathscr{C}C(f, e^{i\theta})$ is also empty and so we have (7.2.5); and if $\mathscr{C}R(f, e^{i\theta})$ contains more than two values, $\chi_A(f, e^{i\theta})$ and $\Phi(f, e^{i\theta})$ are both empty by Theorem 2.20 and (7.2.6) follows.

The counter-example which we gave in Chapter 6 to show that Theorem 6.4 (i) is best possible also shows that Theorem 7.2 (i) above is best possible in the sense that the set $\chi^*(f, e^{i\theta})$ cannot be replaced in (7.2.4) by $\chi(f, e^{i\theta})$.

Theorem 7.2 is a concise statement of a situation which, when $\chi(f, e^{i\theta})$ is unrestricted, is somewhat complicated. We therefore set out the various implications of the theorem in this case.

For a given $a \in \overline{\mathscr{C}R(f, e^{i\theta})} \cap C(f, e^{i\theta})$ the alternative possibilities are tabulated as follows:

(1) $a \in \chi^*(f, e^{i\theta})$. In this case, for all $\eta > 0$ we have

$$a \in \overline{\Gamma}(f, |\theta' - \theta| < \eta).$$

This does not exclude the possibility that at the same time $a \in \Phi(f, e^{i\theta})$.

(2) $a \in \mathscr{C}\chi^*(f, e^{i\theta})$, but $a \in \Phi(f, e^{i\theta})$. In this case we can find η_0 and ϵ_0 such that $U(a, \epsilon_0) \cap \Gamma(f, |\theta' - \theta| < \eta_0)$ is empty. Also $\mathscr{C}R(f, e^{i\theta})$ contains at most two values, one of which must be

a since then $a \in \overline{\mathscr{C}R(f, e^{i\theta})} = \mathscr{C}R(f, e^{i\theta})$. On further analysis we find three possibilities in this case.

(a) There is a set of curves c_n on which $f(z)$ tends uniformly to a as $n \to \infty$ and which converges to an arc $z = e^{i\theta'}$, $|\theta' - \theta| < \eta_1 \leqslant \eta_0$. Since $a \in \mathscr{C}\Gamma(f, |\theta' - \theta| < \eta_1)$, this implies that $\Gamma(f, |\theta' - \theta| < \eta_1)$ is empty.

(b) There are two sets of curves c_n^+ and c_n^- converging respectively to arcs $z = e^{i\theta'}$, $\theta \leqslant \theta' \leqslant \theta + \eta_1$ and $\theta - \eta_1 \leqslant \theta' \leqslant \theta$, on which $f(z)$ tends uniformly to a as $n \to \infty$. Then

$$\Gamma(f, 0 < |\theta' - \theta| < \eta_1) \quad \text{and} \quad U(a, \epsilon_0) \cap \Gamma_P(f, e^{i\theta})$$

are both empty.

(c) There is one set of curves c_n^+ or c_n^- with the property described in (b), and the corresponding set $\Gamma(f, \theta < \theta' < \theta + \eta_1)$ or $\Gamma(f, \theta - \eta_1 < \theta' < \theta)$ is empty. The opposite arc $e^{i\theta'}$,

$$\theta - \eta_1 < \theta' < \theta \quad \text{or} \quad \theta < \theta' < \theta + \eta_1,$$

which we denote by δ, is then a Fatou arc, almost all points of which belong to the set $F(f)$ of Fatou points, and the corresponding set $\Gamma_P(f, \delta)$ lies outside $U(a, \epsilon_0)$. For, consider a point $e^{i\theta'} \in \delta$. Since $\eta_1 \leqslant \eta_0$, we have $a \in \mathscr{C}\chi^*(f, e^{i\theta'})$, so that by Lemma 1,

$$a \in \mathscr{C}C(f, e^{i\theta'}) \cup R(f, e^{i\theta'}).$$

Hence, since $a \in \mathscr{C}R(f, e^{i\theta'})$, we can choose η_2 such that, in $D(e^{i\theta'}, \eta_2)$, $1/(f(z) - a)$ is regular and bounded, so that the arc $z = e^{i\theta''}$, $|\theta'' - \theta'| < \eta_2$, is a Fatou arc. Since $e^{i\theta'}$ is an arbitrary point of the arc δ, our assertion is proved.

Evidently the three subsidiary cases (a), (b) and (c) exhaust the possibilities of case (2) and, by Theorem 7.2 (i), this together with case (1) exhausts all the possibilities when no restriction is postulated for $\chi(f, e^{i\theta})$.

Case (1) is illustrated by the univalent mapping function discussed in Chapter 6. In that example, for any a in the semi-open segment $(1, \tfrac{1}{2}i)$ we have $a \in \mathscr{C}R(f, 1) \subseteq \chi^*(f, 1)$, but $a \notin \chi(f, 1)$. Another example is the modular function $\mu(z)$ when a is one of the omitted values $0, 1, \infty$. But in this case $\Gamma(f, |\theta' - \theta| < \eta)$ consists of these three values for all θ and $\eta > 0$, so that $a \in \chi(f, e^{i\theta})$.

Case $(2)(a)$ is illustrated by an unbounded regular function bounded on a spiral converging to K and tending to infinity on a similar spiral, such as we have discussed in § 4 of Chapter 6.

The pattern of corollaries and collateral results is similar to that arising from Theorem 6.4.

COROLLARY 1. *A necessary and sufficient condition for*

$$\mathscr{C}R(f, e^{i\theta}) \subseteq \chi^*(f, e^{i\theta}) \cup \Phi(f, e^{i\theta}) \qquad (7.2.8)$$

is that $\mathscr{C}R(f, e^{i\theta})$ contain no interior point; and if a value

$$a \in \mathscr{C}R(f, e^{i\theta}) \notin \chi^*(f, e^{i\theta}) \cup \Phi(f, e^{i\theta}),$$

then $\qquad a \in \mathscr{C}C(f, e^{i\theta}) \subseteq \operatorname{Int} \mathscr{C}R(f, e^{i\theta}).$

The condition is necessary since by (7.2.4)

$$\mathscr{C}R(f, e^{i\theta}) \cap C(f, e^{i\theta}) \subseteq \overline{\mathscr{C}R(f, e^{i\theta})} \cap C(f, e^{i\theta})$$
$$\subseteq \chi^*(f, e^{i\theta}) \cup \Phi(f, e^{i\theta}),$$

so that $\quad \mathscr{C}R(f, e^{i\theta}) \cap \mathscr{C}\chi^*(f, e^{i\theta}) \cap \mathscr{C}\Phi(f, e^{i\theta}) \subseteq \mathscr{C}C(f, e^{i\theta}).$

The condition is sufficient since it implies

$$\mathscr{C}R(f, e^{i\theta}) \subseteq \operatorname{Fr} R(f, e^{i\theta}).$$

An equivalent statement of this corollary follows from the fact (compare the corollary to Lemma 3 on p. 126) that $\mathscr{C}C(f, e^{i\theta})$ is empty if and only if $\operatorname{Int} R(f, e^{i\theta})$ is empty, i.e.

A necessary and sufficient condition for (7.2.8) to be satisfied is that $\mathscr{C}C(f, e^{i\theta})$ be empty.

A further group of corollaries is obtained if we impose restrictions upon the set $\chi(f, e^{i\theta})$.

COROLLARY 2. *If $\chi^*(f, e^{i\theta})$ is empty, then $\mathscr{C}R(f, e^{i\theta})$ contains at most two values and* $\quad \mathscr{C}R(f, e^{i\theta}) \subseteq \Phi(f, e^{i\theta});$

and if both $\chi^(f, e^{i\theta})$ and $\Phi(f, e^{i\theta})$ are empty, then $\mathscr{C}R(f, e^{i\theta})$ is empty.*

This follows immediately from Theorem 7.2 (ii), for,

$$\chi(f, e^{i\theta}) \cup \chi_*(f, e^{i\theta}) \subseteq \chi^*(f, e^{i\theta}).$$

More generally, from Theorem 7.2 (ii) we have

COROLLARY 3. *If, for some $\eta > 0$, $\Gamma(f, |\theta' - \theta| < \eta)$ is an isolated set, then $\mathscr{C}R(f, e^{i\theta})$ is an isolated set. In particular, if*

$$\Gamma(f, |\theta' - \theta| < \eta)$$

is a finite set, $\mathscr{C}R(f, e^{i\theta})$ is also a finite set.

We may assume that $\mathscr{C}R(f, e^{i\theta})$ contains more than two values so that $\Phi(f, e^{i\theta})$ is empty, and (7.2.6) is satisfied.

The topological character of the sets $\Phi(f, e^{i\theta})$ and $\Phi(f)$ for any function meromorphic in the disc has been investigated recently by Bagemihl [4].

CHAPTER 8

FURTHER BOUNDARY PROPERTIES OF FUNCTIONS MEROMORPHIC IN THE DISC. CLASSIFICATION OF SINGULARITIES

1. Introduction

In Chapter 2 we developed a series of classical theorems by which the properties of a bounded analytic function were determined by the radial limit values on a set E on $|z| = 1$; we cite, in particular, Theorem 2.5 that a bounded analytic function whose radial limits vanish on a set of positive measure on $|z| = 1$ must be identically zero. It is natural to determine the extent to which such a theorem may be extended, and in § 2 we prove an extension (Theorem 8.1) to unbounded functions due to Privalov, in which radial limits are replaced by angular limits, and which has many uses. The method of proof of Theorem 8.1 may be modified to yield the remarkable theorem of Plessner (Theorem 8.2) which gives a characterization of the distribution on $|z| = 1$ of the singularities of a function meromorphic in $|z| < 1$. Theorem 8.2 still involves an approach 'in angle' to points of $|z| = 1$ and in § 3 we shall examine the nature of the additional restrictions which must be placed on the boundary set in order that the condition of approach 'in angle' may be replaced by radial approach; such conditions are intimately connected with the notion of category, and in § 3 we prove a theorem giving the basic relation. The principal theorem of § 3 contains as special cases various results of Lusin and Wolf in this direction.

In § 6 we give a proof of one of Mergelyan's recent approximation theorems which enables us in § 7 to prove the striking theorem that, on a given set E of first category on $|z| = 1$, the radial behaviour of an analytic function can be made to coincide with that of an arbitrary continuous complex-valued function. By choosing E to be of first category and of measure 2π, we may then show, for example, the existence of a function $f(z)$, analytic in $|z| < 1$, having radial limit 0 almost everywhere on $|z| = 1$;

such a function shows the limitations connected with any generalization of Theorem 2.5 of Chapter 2 which is purely measure-theoretic. In § 8 we consider some of the immediate applications of the fundamental existence theorem (Theorem 8.11).

2. Functions with angular limits. Theorems of Privalov and Plessner

We prove first (Theorem 8.1) a theorem of Privalov [1, 3], which, in virtue of Theorem 3.3, is a natural extension of Theorem 2.5; although the theorem has immediate applications as a uniqueness theorem similar to those of Theorem 2.5, its principal use here will be in the proof of Theorem 8.2 below.

THEOREM 8.1. *Let $f(z)$ be analytic in $|z| < 1$, and let E be a set of positive linear measure on $|z| = 1$ such that, for each point $P \in E$, $f(z)$ tends to 0 in every angle with vertex at P, i.e. P is a Fatou point at which the Fatou value of $f(z)$ is 0. Then $f(z)$ is identically zero in $|z| < 1$.*

Since $m(E) > 0$, it follows from Egorov's theorem† that there exists a perfect, nowhere dense subset E_1 of E, also of positive measure, such that $f(r e^{i\theta})$ tends uniformly to zero as $r \to 1$ inside an angle of opening $\pi/2$ with vertex at $e^{i\theta} \in E_1$ and bisected by the radius drawn to $e^{i\theta}$. If I_n denotes one of the arcs forming the complement of E_1 on $|z| = 1$, we form at each end-point of I_n an angle of opening $\frac{1}{2}\pi$ and bisected by the radius drawn to this end-point. If $m(I_n) < \pi$, two sides of these two angles will intersect inside $|z| < 1$, forming a domain Δ_n not containing $z = 0$, and bounded by I_n and two segments, one emanating from each of the end-points of I_n. In this way there is formed a simply connected domain D in $|z| < 1$, star-shaped with respect to $z = 0$ and bounded either by segments of the sides of the angles having vertices at the end-points of the arcs I_n or by points of E_1. Since the length of the boundary of Δ_n can be shown to be bounded by some fixed multiple of the length of I_n, it follows that the frontier of D consists of a rectifiable Jordan curve intersecting $|z| = 1$ in precisely the set E_1. (We remark that, if one of the arcs I_n has the

† See the footnote on p. 32 concerning the applicability of Egorov's theorem.

property that $m(I_n) \geqslant \pi$, then one may truncate the two angles having vertices at the end-points of I_n by an arc γ_0 of a circle $|z| = r_0$, $0 < r_0 < 1$, to form a domain Δ_0 bounded by γ_0, I_n, and two rectilinear segments; the formation and the properties of the modified domain D are now obvious.)

Let $z = z(w)$, $z(0) = 0$, map $|w| < 1$ conformally onto the domain D; the set E_w on $|w| = 1$, which is the image of E_1 under the mapping $z = z(w)$, has positive linear measure, by Theorem 3.3; furthermore, the function $F(w) = f(z(w))$ is analytic and bounded in $|w| < 1$ and has radial limit 0 on a set of positive measure on $|w| = 1$, so that $F(w)$ and, *a fortiori* $f(z)$, must be identically zero, by Theorem 2.5. This proves Theorem 8.1.

We remark that our proof shows that the restriction in Theorem 8.1 that $f(z)$ should admit 0 as an angular limit may be relaxed to the extent that, for each point of a set E of positive measure on $|z| = 1$, there should exist an angle Δ_θ of opening $\delta > 0$ with vertex at the point $e^{i\theta} \in E$ such that $f(z_n) \to 0$ for all sequences $z_n \to e^{i\theta}$ lying within Δ_θ. Furthermore, the proof indicates that, even if $f(z)$ is meromorphic in $|z| < 1$, rather than analytic, there can be at most a finite number of poles of $f(z)$ inside the domain D, so that an obvious modification of the proof of Theorem 8.1 yields†

COROLLARY 1. *If $f(z)$ is meromorphic in $|z| < 1$, and if, at each point $e^{i\theta}$ belonging to a set E on $|z| = 1$ of positive measure, there exists an angle of opening $\delta > 0$ and bisected by the radius to $e^{i\theta}$ such that* $\lim\limits_{z \to e^{i\theta}} f(z) = 0$ *inside this angular domain, then $f(z)$ is identically zero in $|z| < 1$.*

We remark further that, with no additional restrictions on the set E, the condition that there exists an angle with opening $\delta > 0$ is essential in uniqueness theorems of this type for general meromorphic functions; from Theorem 8.11 below we shall show how it is possible to construct in $|z| < 1$ an example‡ of an analytic function whose radial limit values are zero for almost all points on $|z| = 1$.

Before stating Theorem 8.2, we recall that, if Δ denotes a

† Plessner [1].
‡ Such an example was first given by Lusin and Privalov [1].

domain inside $|z| < 1$, two sides of which consist of distinct recti-linear segments terminating at a point $e^{i\theta}$, then the cluster set of $f(z)$ at $e^{i\theta}$ with respect to the domain Δ will be denoted by $C_\Delta(f, e^{i\theta})$. We shall call a point $e^{i\theta}$ a *Plessner point* for $f(z)$ if $C_\Delta(f, e^{i\theta})$ is total for every angle Δ between pairs of chords through $e^{i\theta}$ however small the angle may be, and the set of Plessner points for $f(z)$ on $|z| = 1$ will be denoted by $I(f)$. We recall that a point $e^{i\theta}$ is called a Fatou point for $f(z)$ if

$$C_{\mathscr{A}}(f, e^{i\theta}) = \bigcup_\Delta C_\Delta(f, e^{i\theta})$$

is degenerate and if $\lim f(z)$ exists uniformly as $z \to e^{i\theta}$ in any angle Δ between chords through $e^{i\theta}$; the set of Fatou points has been denoted by $F(f)$. An arc of $|z| = 1$ in the vicinity of which $|f(z)|$ is uniformly bounded will be called a *Fatou arc* of $f(z)$; it is clear that almost all points of a Fatou arc are Fatou points of $f(z)$.

THEOREM 8.2. *If $f(z)$ is meromorphic in $|z| < 1$, then almost all points of $|z| = 1$ belong to $F(f) \cup I(f)$.*

We denote by E_1 the set of points $e^{i\theta}$ for which $C_\Delta(f, e^{i\theta})$ is degenerate for all Δ with vertex at $e^{i\theta}$, i.e. where $C_\Delta(f, e^{i\theta})$ consists of a single point, and remark that this implies that $\lim\limits_{z \to e^{i\theta}} f(z)$ exists 'in angle' at $e^{i\theta}$; let E_2 denote the set of $e^{i\theta}$ for which $C_\Delta(f, e^{i\theta})$ is total for any Δ with vertex at $e^{i\theta}$; and let E_3 denote the remaining set of points on $|z| = 1$. Since E_1 and E_2 are measurable, it follows that E_3 is measurable, and we make the assumption that $m(E_3) > 0$. Now the set of all curvilinear triangles on the sphere whose vertices have rational coordinates is countable; we denote the set of such triangles by $\tau_1, \tau_2, ..., \tau_n,$ Then, if $e^{i\theta_0}$ does not belong to E_2, we can find an angular region Δ with vertex at $e^{i\theta_0}$ such that $C_\Delta(f, e^{i\theta_0})$ is not total, so that there exists at least one triangle τ_k such that $\tau_k \cap C_\Delta(f, e^{i\theta}) = \varnothing$.

Inside Δ we may choose another angular region Δ_1 in the form of a circular sector with vertex $e^{i\theta_0}$ and with radius $1/m$, where m is an integer, with the property that in Δ_1 the function $f(z)$ assumes no value of τ_k. Furthermore, we may take the angular opening of Δ_1 to be $\pi/2^n$, where n is another integer and where the angle bisector of Δ_1 at $e^{i\theta_0}$ forms with the radius of $|z| < 1$ drawn to $e^{i\theta_0}$ an angle $p\pi/q$ which is a rational multiple of π.

We denote by $E_{n,m,p,q,k}$ the set of all $e^{i\theta} \in E_3$ for which the above constructions are possible; we remark that the class of these sets is countable since the five indices are taken over the natural numbers. The measurability of the sets $E_{n,m,p,q,k}$ follows at once from the fact that they are defined by means of inequalities, equalities and limiting processes applied to a measurable function. Now if $m(E_3) > 0$, then at least one of these sets, say $E_{n_0 m_0 p_0 q_0 k_0}$, is of positive measure and contains a perfect subset E' which is also of positive measure. If w_0 is an interior point of τ_{k_0}, there exists a positive number $\delta > 0$ such that

$$\frac{1}{|f(z) - w_0|} < \delta$$

in those regions Δ_1 with vertices at points $e^{i\theta} \in E'$. We extend each of the sides of Δ_1 from the point $e^{i\theta} \in E'$ until it either meets a side of another Δ_1 or else meets the circle $|z| = \rho_0 > 1 - 1/m_0$. The union of the circle $|z| < \rho_0$ and the interiors of the domains Δ_1 forms a simply connected domain D_z whose frontier intersects $|z| = 1$ in the set E'. It follows from the construction of D_z that $\operatorname{Fr} D_z$ is a rectifiable Jordan curve.

The function $[f(z) - w_0]^{-1}$ can have at most a finite number of poles a_1, a_2, \ldots, a_ν in D_w; since these poles must lie in $|z| \leqslant \rho_0$, the function $F(z) = (z - a_1)(z - a_2) \cdots (z - a_\nu)[f(z) - w_0]^{-1}$ is then analytic and bounded in D_z, and, by Theorem 3.3, must possess a unique limiting value as z approaches a point of $\operatorname{Fr} D_z$ for almost all points of $\operatorname{Fr} D_z$. Since $\operatorname{Fr} D_z$ has a tangent at almost all points of E' and since such a tangent must coincide with the tangent to $|z| = 1$, whenever it exists, it follows that there exists at least one point $e^{i\theta_0}$ of E for which $F(z)$, and $a\ fortiori\ f(z)$, possesses a unique limiting value in any angle of opening less than π and bisected by the radius $\theta = \theta_0$. This shows that $e^{i\theta_0}$ belongs to E_1 and not to E_3 as we assumed, so that we have forced the contradiction which proves Theorem 8.2.

By means of an appropriate conformal mapping, it can be seen that, if a function $f(z)$ is meromorphic in a domain D bounded by a rectifiable Jordan curve C, then C may be decomposed into three sets E_1, E_2, and E_3, relative to which $f(z)$ has the behaviour

described in Theorem 8.2. As a particularly useful corollary to Theorem 8.2, we have

COROLLARY 2. *If $f(z) = u(z) + iv(z)$ is meromorphic in $|z| < 1$ and if, on some set E of positive measure on $|z| = 1$, one of the four conditions*

$$\overline{\lim_{z \to e^{i\theta}}} u(z) < \infty, \quad \varliminf_{z \to e^{i\theta}} u(z) > -\infty, \quad \overline{\lim_{z \to e^{i\theta}}} v(z) < \infty, \quad \varliminf_{z \to e^{i\theta}} v(z) > -\infty$$

is satisfied, where $z \to e^{i\theta}$ inside some angle of fixed opening with vertex at $e^{i\theta}$, then $f(z)$ possesses angular limits at almost all points of E.

3. Radial cluster sets and uniqueness theorems

We have already defined two classes of boundary points for functions meromorphic in $|z| < 1$ in terms of cluster sets, namely *Fatou points* and *Plessner points*. We now introduce another.

A point $e^{i\theta}$ on $K: |z| = 1$ such that the complete cluster set $C(f, e^{i\theta})$ is total will be called a *Weierstrass point* of $f(z)$, and we shall denote the set of all such points on K by $W(f)$. Given an open arc $A \subseteq K$, we shall write $W(f, A)$ for $W(f) \cap A$, and, as usual, $\mathscr{C}W(f) = K - W(f)$.

We suppose now that $e^{i\theta} \in \mathscr{C}W(f, A)$; then $\mathscr{C}C(f, e^{i\theta})$ is not empty and thus is open and contains a circular disc of centre a, say, and (spherical) radius $\epsilon > 0$. There is therefore a number $\delta > 0$ such that, for

$$z \in d = \{z : |z| < 1, |z - e^{i\theta}| < \delta\},$$

$f(z)$ does not meet the disc of centre a and radius $\frac{1}{2}\epsilon$. The arc $K \cap \bar{d}$, which by choice of ϵ is contained in A, thus contains an open Fatou arc of $f(z)$, and we have proved

LEMMA 1. *If an arc A of K contains no Fatou arc of a function $f(z)$, meromorphic in $|z| < 1$, then $A \subset W(f)$.*

We denote by $E(f)$ the set of points $e^{i\theta} \in K$ such that $C_\rho(f, e^{i\theta})$ is subtotal, and we prove

THEOREM 8.3. *Let $f(z)$ be meromorphic in $|z| < 1$ and let $E(f)$ be of second category on some arc $A \subset K$. If*

(i) *there is a number b and a set M which is metrically dense†
on A such that* $b \in \bigcap_{e^{i\theta} \in M} C_\rho(f, e^{i\theta})$, *then* $f(z) \equiv b$; *while if*

(ii) $f(z)$ *is not a constant, then the set* $\Gamma_P(f, A)$ *is of positive
linear measure.*

It follows from Theorem 4.8, by taking \mathscr{G}_θ to be the radius
(or any other curve of fixed form, such as a chord of fixed direc-
tion relative to the radius), that $E(f)$ differs from $\mathscr{C}W(f)$ by at
most a set of first category. Since by hypothesis $E(f) \cap A$ is of
second category, it follows that the same is true of $\mathscr{C}W(f) \cap A$,
which is therefore not empty. By Lemma 1, A therefore contains
a Fatou arc B. It follows from Theorem 2.4 that M contains a
subset of $F(f)$ metrically dense on B, so that (i) now follows from
Theorem 2.5. To prove (ii) we observe that if $f(z)$ is not constant,
Theorem 2.5 ensures the existence of points $e^{i\theta_1}$, $e^{i\theta_2}$ on B such
that $f(e^{i\theta_1}) \neq f(e^{i\theta_2})$. If we apply Lemma 1 of Chapter 6, (ii) now
follows, and the theorem is proved.

Theorem 8.3 (i) yields as immediate corollaries two classical
theorems concerning the uniqueness of analytic functions:

COROLLARY 3.‡ *If* $f(z)$ *is analytic in* $|z| < 1$ *and has the same
radial limit a, i.e.* $C_\rho(f, e^{i\theta}) = a$, *on a set M which is both metrically
dense and of second category on an arc of* $|z| = 1$, *then* $f(z) \equiv a$.

COROLLARY 4.§ *Let* $f(z)$ *be analytic in* $|z| < 1$, *and let M be a set
of type* G_δ *which is dense on an arc A of* $|z| = 1$ *and such that
$\infty \in \mathscr{C}C_\rho(f, e^{i\theta})$ for all $e^{i\theta} \in M$. Then, if there is a finite number a such
that* $a \in C_\rho(f, e^{i\theta})$ *almost everywhere on A, the function* $f(z)$ *is
identically constant.*

We have only to note that a set of type G_δ which is dense on A
is residual on A.

For other results in the direction of this paragraph, we refer to
the papers of Collingwood [1] and Bagemihl and Seidel [1].

4. Classification and distribution of singularities on K

Denote by $F_1(f)$ the set of points $e^{i\theta} \in K$ at which the set $F(f)$
of Fatou points of $f(z)$ is of density 1. It is implicit in the proof of

† I.e. for every sub-interval $\alpha \subseteq A$ the measure $m(M \cap \alpha) > 0$. Cf. Colling-
wood [1], p. 169. ‡ Lusin and Privalov [1]. § Wolf [1].

Lemma 1 in § 3 above that if $e^{i\theta} \in \mathscr{C}W(f)$, then $e^{i\theta} \in F_1(f)$, and we have

THEOREM 8.4. *If $f(z)$ is meromorphic in $|z| < 1$, then every point $e^{i\theta}$ belongs either to $W(f)$ or to $F_1(f)$.*

This yields

COROLLARY 5. *If $f(z)$ is meromorphic in $|z| < 1$, then almost all points $e^{i\theta}$ belong either to $W(f)$ or to $F(f)$.*

For, if $\mathscr{C}W(f) \cap \mathscr{C}F(f)$ is of positive measure there is a point $e^{i\theta_0}$, say, at which this set is of density 1 and at which therefore $\mathscr{C}F(f)$ is of density 1. But $e^{i\theta_0} \in \mathscr{C}W(f) \subset F_1(f)$, so that we have a contradiction.

We note that, since $I(f) \subset W(f)$, Corollary 5 is also a corollary of the deeper Theorem 8.2.

It follows from Lemma 1 of Chapter 6 that if $\Gamma_F(f)$ is of linear measure zero, K contains no Fatou arc, so that, by Lemma 1 of § 3 above, $K \subset W(f)$ and, as a counterpart to Theorem 8.4, we have proved

THEOREM 8.5. *If $f(z)$ is meromorphic in $|z| < 1$ and if the set $\Gamma_F(f)$ of Fatou values of $f(z)$ is of linear measure zero, then $K \subset W(f)$.*

Exactly as in the proof of Theorem 4.2, it follows from the definition that $\mathscr{C}R(f, e^{i\theta})$ is of first category for all $e^{i\theta} \in W(f)$. Subsets of $W(f)$ for which significant theorems may be proved are defined by imposing further restrictions on $\mathscr{C}R(f, e^{i\theta})$. Thus, a classical theorem of Frostman which states that a function φ for which $\mathscr{C}R(\varphi)$ is of positive capacity has bounded characteristic $T(r, \varphi)$ enables us, by similar arguments, to extend Theorem 8.4 and Corollary 5 to the set $H(f)$ of points $e^{i\theta}$ for which $\mathscr{C}R(f, e^{i\theta})$ is of capacity zero. By analogy, we call these *Frostman points.*†
The analogue for the set $H(f)$ of Theorem 8.5, which we state without proof, is the following:‡

(*) *If $f(z)$ is meromorphic in $|z| < 1$ and if $\Gamma_F(f)$ is of capacity zero, then $K \subset H(f)$.*

† This is the terminology adopted by Collingwood and Cartwright [1], p. 138.
‡ For proofs see Collingwood and Lohwater [2], p. 100, and Cartwright and Collingwood [1].

The set $P(f)$ of *Picard points* has already been defined (Chapter 2, § 9). The classification may be extended in the following way. Denote by $P_n(f)$, $0 \leqslant n < \infty$, the set of points $e^{i\theta} \epsilon K$ such that $\mathscr{C}R(f, e^{i\theta})$ contains at most n values, and by $P_{\aleph_0}(f)$ the set for which $\mathscr{C}R(f, e^{i\theta})$ is at most countable. Then, plainly, for any finite $n \geqslant 0$, $P_n(f) \subset P_{n+1}(f) \subset P_{\aleph_0}(f)$ and, in particular,
$$P(f) = P_2(f) \supset P_0(f) \cup P_1(f).$$

THEOREM 8.6. *If* $f(z)$ *is meromorphic in* $|z| < 1$ *and if*

(i) *for some finite* $n \geqslant 0$ *the set* $\Gamma_F(f)$ *contains at most* n *values,*
then
$$K \subset P(f) \cup P_n(f);$$
while if

(ii) $\Gamma_F(f)$ *is at most countable, then*

$$K \subset P_{\aleph_0}(f).$$

To prove this theorem, we suppose that $e^{i\theta}$ is not a point of $P(f)$. We can therefore find $\eta > 0$ such that in the subdomain $V(\theta, \eta)$ of $|z| < 1$ bounded by the arc $(e^{i(\theta-\eta)}, e^{i(\theta+\eta)})$ containing $e^{i\theta}$ and the circular arc of centre $e^{i\theta}$ passing through the points $e^{i(\theta-\eta)}$ and $e^{i(\theta+\eta)}$ the function $f(z)$ omits at least three values. Under a conformal mapping of $V(\theta, \eta)$ onto the unit disc the sets $\Gamma(f, |\theta' - \theta| < \eta)$ and $\Gamma_P(f, |\theta' - \theta| < \eta)$ defined in § 1 of Chapter 7 are invariant on the image arc, and it follows from Theorem 2.20 and the remarks following Theorem 6.5 on p. 130 that

$$\Gamma(f, |\theta' - \theta| < \eta) = \Gamma_P(f, |\theta' - \theta| < \eta) = \Gamma_F(f, |\theta' - \theta| < \eta)$$

and hence that
$$\chi(f, e^{i\theta}) = \chi_F(f, e^{i\theta}), \tag{8.4.1}$$

where $\chi_F(f, e^{i\theta}) = \bigcap_\eta \Gamma_F(f, |\theta' - \theta| < \eta)$ and $\Gamma_F(f, |\theta' - \theta| < \eta)$ is the set of Fatou values attained on the arc $(e^{i(\theta-\eta)}, e^{i(\theta+\eta)})$.

Now, by hypothesis, $\Gamma_F(f, |\theta' - \theta| < \eta) \subset \Gamma_F(f)$ is either finite or countable and thus of linear measure zero. Theorem 7.2 (ii) is therefore applicable and, by (7.2.6) and (8.4.1),

$$\mathscr{C}R(f, e^{i\theta}) \subset \chi_F(f, e^{i\theta}) \subset \Gamma_F(f).$$

Since $e^{i\theta} \epsilon \mathscr{C}P(f)$, this proves both parts of the theorem.

The above argument also yields very easily the following theorem, in which $F'(f)$ denotes the derived set of $F(f)$.

THEOREM 8.7. *If f(z) is meromorphic in* $|z| < 1$, *then*

$$K \subset F'(f) \cup P(f).$$

For if $e^{i\theta} \in \mathscr{C}P(f)$, (8.4.1) holds so that

$$\chi^*(f, e^{i\theta}) = \bigcap_\eta \overline{\Gamma}_F(f, |\theta' - \theta| < \eta). \qquad (8.4.2)$$

Furthermore, $\Phi(f, e^{i\theta})$ is empty, so that (7.2.4) and (8.4.2) imply

$$\mathscr{C}R(f, e^{i\theta}) \subset \bigcap_\eta \overline{\Gamma}_F(f, |\theta' - \theta| < \eta). \qquad (8.4.3)$$

But $\mathscr{C}R(f, e^{i\theta})$ contains at least three values, and it follows from (8.4.3) that, for arbitrarily small η, the arc $(e^{i(\theta-\eta)}, e^{i(\theta+\eta)})$ contains points of $F(f)$, so that $e^{i\theta} \in F'(f)$.

The theorems of this paragraph have shown that there is an intimate relation between the distribution of the Fatou points on K and the size of the set Γ_F of Fatou values of a meromorphic function on the one hand and the nature and distribution of its singularities on the other. In particular, writing $m\Gamma$ for the linear measure and cap Γ for the capacity of a set Γ, the following table sets out the implications of Theorems 8.5, (*) and 8.6:

$$m\Gamma_F(f) = 0 \Rightarrow K \subset W(f);$$
$$\text{cap } \Gamma_F(f) = 0 \Rightarrow K \subset H(f);$$
$$\Gamma_F(f) \text{ countable} \Rightarrow K \subset P_{\aleph_0}(f);$$
$$\Gamma_F(f) \text{ contains at most } n \text{ values} \Rightarrow \begin{cases} K \subset P_n(f) & \text{if } 2 < n < \infty \\ K \subset P(f) & \text{if } 0 \leqslant n \leqslant 2. \end{cases}$$

5. Meier's analogue of Plessner's Theorem

A theorem similar to Plessner's theorem, but in terms of category instead of measure, was proved by Meier [3]. In the statement of the theorem the set of Fatou points is replaced by the set $M(f)$ of *Meier points* defined as follows. The point $e^{i\theta} \in M(f)$ if (i) $C(f, e^{i\theta})$ is subtotal, and (ii) the chordal cluster set $C_{\rho(\varphi)}(f, e^{i\theta}) = C(f, e^{i\theta})$ for all values of φ in $[-\frac{1}{2}\pi, \frac{1}{2}\pi]$. We note that there is no relation of inclusion between $F(f)$ and $M(f)$. Meier's theorem is the following.

THEOREM 8.8. *If $f(z)$ is meromorphic in $|z| < 1$, then all points of $|z| = 1$ except a set of first category belong to $M(f) \cup I(f)$.*

There are two cases to consider: (i) in which the global cluster set $C(f)$ is subtotal, and (ii) in which it is total.

We start with (i). A transformation will make the function bounded and holomorphic and we shall assume this. $I(f)$ is then empty and we have to prove

THEOREM 8.9. *If $f(z)$ is analytic and bounded in $|z| < 1$, then all points of $|z| = 1$ except a set of first category belong to $M(f)$.*

This is an analogue of Fatou's theorem in terms of category. To prove it, let E be the set of points $e^{i\theta} \in \mathscr{C}M(f)$, i.e. such that there is some chord $\rho(\varphi)$ for which $C_{\rho(\varphi)}(f, e^{i\theta}) \neq C(f, e^{i\theta})$. Write $P = e^{i\theta}$ and let $P \in E$ so that $C_{\rho(\varphi)}(f, P) \neq C(f, P)$ for some φ in $[-\frac{1}{2}\pi, \frac{1}{2}\pi]$. We choose $a \in C(f, P) \cap \mathscr{C}C_{\rho(\varphi)}(f, P)$. There is then some ϵ-neighbourhood $\eta(a)$ of a in $C(f, P) \cap C_{\rho(\varphi)}(f, P)$ and we can find R_0 such that $f(z)$ takes no value in $\eta(a)$ on the terminal segment of $\rho(\varphi)$ of length R_0 which we call the R_0-segment. Now, given ξ on this R_0-segment it follows from elementary geometry that the disc $\chi(R, \xi)$ defined by $|z - \xi| < R \sin(\frac{1}{4}\pi - \frac{1}{2}\varphi)$, where $R = |\xi - P|$, lies inside the unit disc so that f is bounded (we assume $|f| < 1$) in $\chi(R, \xi)$. We now apply Schwarz's lemma which gives

$$|f(z) - f(\xi)| < \frac{2|z - \xi|}{R \sin(\frac{1}{4}\pi - \frac{1}{2}\varphi)} \qquad (8.5.1)$$

for z in $\chi(R, \xi)$. Putting $\sigma(R) = \frac{1}{4}\epsilon R \sin(\frac{1}{4}\pi - \frac{1}{2}\varphi)$ it follows from (8.5.1) that

$$|f(z) - f(\xi)| < \tfrac{1}{2}\epsilon \qquad (8.5.2)$$

for

$$|z - \xi| < \sigma(R),$$

while

$$|f(\xi) - a| > \epsilon.$$

Hence

$$|f(z) - a| > \tfrac{1}{2}\epsilon \qquad (8.5.3)$$

for

$$|z - \xi| < \sigma(R).$$

But the sides of the angle of magnitude $\frac{1}{2}\pi - \varphi$ bisected by $\rho(\varphi)$ are tangent to the perimeters of all the discs $\chi(R, \xi)$ so that in any smaller angle contained in this one (8.5.3) is satisfied. Since $a \in C(f, P)$ it now follows from Theorem 4.10 that E is of first category, which proves the theorem.

We prove Theorem 8.8 by a reduction to the conditions of Theorem 8.9. Let E be the set of points of $|z| = 1$ not belonging to $I(f)$. This is the set $E_1 \cup E_3$ defined in the proof of Theorem 8.2. We use the same technique and notation here, but now applied to E on the assumption that E is of second category, as was applied there to E_3 on the assumption that $m(E_3) > 0$. Thus, we denote by $E_{n,m,p,q,k}$ the set of points $e^{i\theta}$ such that $f(z)$ takes no value in the triangle τ_k in a circular sector of vertex $e^{i\theta}$ whose magnitude and orientation are defined by the integers n, m, p, q. The class of sets $E_{n,m,p,q,k}$ is countable since the five indices are taken over the natural numbers and plainly the set E is their union. If E is of first category there is nothing to prove; but if E is of second category one or more (and perhaps a countable infinity) of these subsets of E are of second category. We denote any one of them by X, its closure by \bar{X} and the set of its frontier points by Y. Since $\bar{X} - Y$ is open it consists of a countable set of intervals J_ν on each of which the points of X are dense and the corresponding set of sectors covers an annular trapezoid V_ν in which, for some $a_K \in \tau_K$, K being the fifth index of X, and some $\epsilon > 0$, $f(z)$ satisfies $|f(z) - a| > \epsilon$. The function $F(z) = 1/(f(z) - a_K)$ is therefore bounded in V_ν. We may transform V_ν to the disc and apply Theorem 8.9 which shows that all points of $X \cap J_\nu$ except for a set of first category belong to $M(F)$ and therefore to $M(f)$. Since $\bar{X} = Y + \cup J_\nu$, Y is non-dense and X belongs to a countable class of sets, this proves the theorem.

6. A theorem on polynomial approximation

The theorem of Weierstrass concerning the approximation by polynomials of a function continuous on an interval has led, in a natural way, to an analysis of the nature of those sets E in the plane, more general than intervals, on which a given continuous function may be approximated uniformly by a sequence of polynomials or rational functions. With no other restrictions than continuity of the given function f, it is clear that the closure of the set E cannot contain interior points, for, if E were to contain an open disc γ, then f, as the uniform limit of a sequence of analytic functions, must be analytic in γ. Furthermore, the set E cannot separate the plane, for otherwise, there would exist a

bounded component of the complement of E whose frontier lies wholly in E so that the uniform limit f of the sequence could then be continued analytically into this component, a fact which is not generally valid for an arbitrary continuous function on E.

Since an extended account of the theory of such approximations is well outside the scope of this monograph, we shall restrict ourselves to a recent theorem of Mergelyan† which is essential to the developments of § 7 below.

THEOREM 8.10. *Let $f(z)$ be continuous on a compact set E, and analytic at interior points of E, if such exist. Then a sufficient condition that there exists a sequence of polynomials converging uniformly to $f(z)$ on E is that E does not separate the plane.*

Since E is closed, we may find a function $\varphi(z)$ which is continuous in the whole plane and which coincides with $f(z)$ on the set E.‡ Let $R_0 > 1$ be a real number such that E is contained in the circle $|z| < R_0$, and let $\omega(\delta)$ be the modulus of continuity of $\varphi(z)$ in $|z| < R_0$, i.e. the least upper bound of $|\varphi(z_1) - \varphi(z_2)|$ under the conditions $|z_1| \leqslant R_0$, $|z_2| \leqslant R_0$, $|z_1 - z_2| \leqslant \delta$. We denote by H_δ the set of all points of E whose distance from FR E exceeds δ, $0 < \delta < 1$, and by R_δ the set $E - H_\delta$; the set H_δ may, of course, be empty.

We may exhaust the complement of E by means of a sequence of domains G_n with $G_n \subset G_{n+1}$ and such that FR G_n consists of a finite number of analytic Jordan curves. We denote by $R_{n\delta}$ the set $\mathscr{C}G_n - H_\delta$ and define the functions

$$K(r) = \begin{cases} \dfrac{3}{\pi\delta^2}\left(1 - \dfrac{r}{\delta}\right), & 0 < r \leqslant \delta, \\ 0, & r > \delta; \end{cases} \qquad (8.6.1)$$

$$\Phi(z) = \iint_{|\zeta| < 2R_0} \varphi(\zeta)\, K(|\zeta - z|)\, d\xi\, d\eta, \qquad (8.6.2)$$

where $\zeta = \xi + i\eta$. It follows immediately from the definition (8.6.1) that

$$\iint_{|\zeta| < 2R_0} K(|\zeta - z|)\, d\xi\, d\eta = 1$$

† See Mergelyan [1], where the problem is treated completely.
‡ See, for example, Alexandroff and Hopf [1], pp. 73 ff.

for $z \in E$. Furthermore,

$$\iint_{|\zeta|<2R_0} \frac{\partial}{\partial x} K(|\zeta-z|)\,d\xi\,d\eta = \iint_{|\zeta-z|\leqslant\delta} \frac{\partial K}{\partial r}\frac{(x-\xi)}{|z-\zeta|}\,d\xi\,d\eta = 0,$$

while
$$\iint_{|\zeta|<2R_0} \left|\frac{\partial K}{\partial x}\right|\,d\xi\,d\eta = \iint_{|\zeta-z|\leqslant\delta} \left|\frac{\partial K}{\partial r}\right|\frac{|x-\xi|}{|\zeta-z|}\,d\xi\,d\eta$$
$$\leqslant \frac{3}{\pi\delta^3}\iint_{|z-\zeta|\leqslant\delta} d\xi\,d\eta \leqslant \frac{3}{\delta}.$$

Similarly,

$$\iint_{|\zeta|<2R_0} \frac{\partial}{\partial y} K(|\zeta-z|)\,d\xi\,d\eta = 0 \quad (z \in E),$$

and
$$\iint_{|\zeta|<2R_0} \left|\frac{\partial K}{\partial y}\right|\,d\xi\,d\eta < \frac{3}{\delta}.$$

Now

$$|\Phi(z)-\varphi(z)| \leqslant \sup_{|z-\zeta|\leqslant\delta}|\varphi(\zeta)-\varphi(z)|\iint_{|z-\zeta|\leqslant\delta} K(|z-\zeta|)\,d\xi\,d\eta = \omega(\delta)$$

for $|z| < R_0$, while $\Phi(z)-\varphi(z) = 0$ for $z \in H_\delta$. If $\Phi(z) = U(z)+iV(z)$, then

$$\frac{\partial U}{\partial x}+i\frac{\partial V}{\partial x} = \iint_{|\zeta|<2R_0} [u(\zeta)+iv(\zeta)]\frac{\partial K}{\partial x}\,d\xi\,d\eta,$$

whence
$$\left|\frac{\partial U}{\partial x}\right| = \left|\iint_{|\zeta|<2R_0} u(\zeta)\frac{\partial K}{\partial x}\,d\xi\,d\eta\right|$$
$$= \left|\iint_{|\zeta|<2R_0} [u(\zeta)-u(z)]\frac{\partial K}{\partial x}\,d\xi\,d\eta\right| \leqslant \frac{3\omega(\delta)}{\delta}.$$

Identical bounds are obtained for

$$\left|\frac{\partial U}{\partial y}\right|, \quad \left|\frac{\partial V}{\partial x}\right| \quad \text{and} \quad \left|\frac{\partial V}{\partial y}\right|,$$

so that
$$\left|\left(\frac{\partial U}{\partial x}-\frac{\partial V}{\partial y}\right)+i\left(\frac{\partial U}{\partial y}+\frac{\partial V}{\partial x}\right)\right| \leqslant \frac{12\omega(\delta)}{\delta}. \tag{8.6.3}$$

Now if a function $g(z)$ is analytic in a domain D having a rectifiable boundary γ, and if $h(z) = h_1(z)+ih_2(z)$ is continuously differentiable in $D \cup \gamma$, then Green's formula, as applied to the real and imaginary parts of the line integral

$$\int_\gamma g(\zeta)h(\zeta)\,d\zeta,$$

yields the identity

$$\int_\gamma g(\zeta)h(\zeta)\,d\zeta = \iint_D g(\zeta)\left\{\frac{\partial h_1}{\partial \eta} - \frac{\partial h_2}{\partial \xi} + i\left(\frac{\partial h_2}{\partial \eta} - \frac{\partial h_1}{\partial \xi}\right)\right\}d\xi\,d\eta.$$

If we apply this formula to the functions $1/(\zeta - z)$ and $\Phi(z)$ in the domain obtained by deleting from D a small circle of radius ϵ with centre z and allow ϵ to approach zero, we obtain the identity

$$\Phi(z) = \frac{1}{2\pi i}\int_\gamma \frac{\Phi(\zeta)}{\zeta - z}\,d\zeta + \frac{1}{2\pi}\iint_D \frac{\left(\dfrac{\partial U}{\partial \xi} - \dfrac{\partial V}{\partial \eta}\right) + i\left(\dfrac{\partial U}{\partial \eta} + \dfrac{\partial V}{\partial \xi}\right)}{\zeta - z}\,d\xi\,d\eta, \tag{8.6.4}$$

which is an extension of Cauchy's integral formula to differentiable, but not necessarily analytic, functions.

It now follows from (8.6.4), as applied to the open set

$$\mathscr{G}_n = \mathscr{C}G_n - \mathrm{Fr}\,G_n$$

that

$$\Phi(z) = \frac{1}{2\pi i}\int_{\mathrm{Fr}\,G_n}\frac{\Phi(\zeta)}{\zeta - z}\,d\zeta + \frac{1}{2\pi i}\int_{\mathrm{Fr}\,H_\delta}\frac{\Phi(\zeta)}{\zeta - z}\,d\zeta$$

$$+ \frac{1}{2\pi}\iint_{R_{n\delta}}\frac{\left(\dfrac{\partial U}{\partial \xi} - \dfrac{\partial V}{\partial \eta}\right) + i\left(\dfrac{\partial U}{\partial \eta} + \dfrac{\partial V}{\partial \xi}\right)}{\zeta - z}\,d\xi\,d\eta, \quad (8.6.5)$$

where we remark that \mathscr{G}_n consists of finitely many components. Now $\Phi(z)$ coincides with $\varphi(z)$ at all points of H_δ, so that, for $z \in \mathrm{Fr}\,E$, the function $\Phi(\zeta)/(\zeta - z)$ is an analytic function of ζ for $\zeta \in H_\delta$, whence the second integral in (8.6.5) vanishes for $z \in \mathrm{Fr}\,E$. Since the first integral in (8.6.5) is an analytic function for $z \in E$, it follows from Runge's theorem† that there exists a polynomial $P_1(z)$ such that the difference

$$\left|P_1(z) - \frac{1}{2\pi i}\int_{\mathrm{Fr}\,G_n}\frac{\Phi(\zeta)}{\zeta - z}\,d\zeta\right| \tag{8.6.6}$$

is arbitrarily small for all $z \in E$. To complete the proof of the theorem, it suffices, in view of the uniformity on E of the limit

$$\lim_{n\to\infty}\frac{1}{2\pi}\iint_{R_{n\delta}}\frac{\Psi(\zeta)}{\zeta - z}\,d\xi\,d\eta = \frac{1}{2\pi}\iint_{E - H_\delta}\frac{\Psi(\zeta)}{\zeta - z}\,d\xi\,d\eta,$$

† Saks and Zygmund [1], p. 177.

where $$\Psi'(\zeta) = \frac{\partial U}{\partial \xi} - \frac{\partial V}{\partial \eta} + i\left(\frac{\partial U}{\partial \eta} + \frac{\partial V}{\partial \xi}\right),$$

to show that the limit as $\delta \to 0$ of the infimum of

$$\left| \frac{1}{2\pi} \iint_{E-H_\delta} \frac{\Psi'(\zeta)}{\zeta - z} d\xi \, d\eta - \mathscr{P}(z) \right|,$$

taken over all polynomials $\mathscr{P}(z)$ in z, is zero. To prove this we need the following

LEMMA. *For any point $\zeta \in R_\delta$ there exists a polynomial $P_\zeta(z)$ in z and two constants c_1 and c_2 such that*

$$|P_\zeta(z)| < \frac{c_1}{\delta} \quad for \quad z \in E, \tag{8.6.7}$$

and

$$\left| P_\zeta(z) - \frac{1}{\zeta - z} \right| < \frac{c_2 \delta^2}{|\zeta - z|^3} \quad for \quad z \in E \quad and \quad |\zeta - z| > 10\delta, \tag{8.6.8}$$

where the degree of $P_\zeta(z)$ is bounded by a number which does not depend on ζ.

We may assume, without loss of generality, that $\zeta = 0$. Since E does not separate the plane, it follows from the definition of R_δ that there exists in $\mathscr{C}E$ an arc l with end-points on the circles $|z| = 2\delta$ and $|z| = 3\delta$ and an analytic Jordan curve γ_0 lying in $\mathscr{C}E \cap \{|z| < 4\delta\}$ whose interior D_z contains the arc l. Let the exterior domain determined by the Jordan curve γ_0 be mapped conformally onto $|w| > 4\delta$ by the function $z = z(w)$, $z(\infty) = \infty$, and let D_w denote the image of D_z under the elementary mapping $w = z/4\delta$. Then the function

$$F(w) = \frac{1}{4\delta} z(4\delta w) = aw + b + \frac{a_1}{w} + \frac{a_2}{w^2} + \cdots$$

maps $|w| > 1$ onto $\mathscr{C}\bar{D}_w$. Now the domain D_w lies in $|w| < 1$ and its diameter exceeds $\frac{1}{4}$, so that there exist constants c_3 and c_4 such that

$$0 < c_3 < |a| < c_4 < \infty.$$

Since $|F(w)| < 1$ on $|w| = 1$, it follows that the function $F(w) - aw$

is analytic and bounded in modulus by $1 + c_4$ in $|w| > 1$, whence $|b| < c_5$, $|a_k| < c_5$ $(k = 1, 2, \ldots)$, so that

$$|F(w) - aw - b| < \frac{4c_5}{|w|} \quad \text{for} \quad |w| > 2.$$

Thus, for the function $z(w)$, we have

$$|z(w) - 4\delta b - aw| < \frac{128\delta^2}{|w|}$$

for $|w| > 8\delta$, and, if $w = w(z)$ denotes the inverse function of $z(w)$, we have

$$\left| \frac{1}{z - 4\delta b} - \frac{1}{aw(z)} \right| < \frac{c_6}{|z - 4\delta b| \cdot |z|^2} < \frac{c_7}{|z|^3} \quad (|z| > 10\delta).$$

On the set E the function $1/[aw(z)]$ is analytic and bounded by c_8/δ, where c_8 is some constant, so that, by Runge's theorem, there exists a polynomial $Q(z)$ with the following properties:

$$\left| \frac{1}{z - 4\delta b} - Q(z) \right| < \frac{c_9 \delta^2}{|z|^3} \quad \text{for} \quad z \in E \quad \text{and} \quad |z| > 10\delta; \quad (8.6.9)$$

$$|Q(z)| < \frac{c_{10}}{\delta} \quad (z \in E); \quad (8.6.10)$$

$$|Q(z)| < \frac{c_{11}}{|z|} \quad (z \in E). \quad (8.6.11)$$

Thus, for $z \in E$ and $|z| > 10\delta$, we have

$$\left| \frac{4\delta b}{(z - 4\delta b)^2} - 4\delta b Q^2 \right| = 4\delta |b| \left| \frac{1}{z - 4\delta b} - Q \right| \left| \frac{1}{z - 4\delta b} + Q \right| \leqslant \frac{4\delta |b| c_9}{|z|^3} \frac{c_{12}}{|z|},$$

whence $\quad \left| \dfrac{4\delta b}{(z - 4\delta b)^2} - 4\delta b Q^2 \right| < \dfrac{c_{13}}{|z|^3} \quad (z \in E, |z| > 10\delta), \quad (8.6.12)$

where c_{13} is another constant. Applying (8.6.9) and (8.6.12) to the identity

$$\frac{1}{z} = \frac{1}{z - 4\delta b} - \frac{4\delta b}{(z - 4\delta b)^2} + \frac{16\delta^2 b^2}{z(z - 4\delta b)^2},$$

we have, for $z \in E$ and $|z| > 10\delta$,

$$\left| \frac{1}{z} - (Q - 4\delta b Q^2) \right| < \frac{c_9 \delta^2}{|z|^3} + \frac{c_{13} \delta^2}{|z|^3} + \frac{64\delta^2}{|z| |z - 4\delta b|^2},$$

so that $\quad \left| \dfrac{1}{z} - (Q - 4\delta b Q^2) \right| < \dfrac{c_{14}\delta^2}{|z|^3} \quad (z \in E, \ |z| > 10\delta),$ (8.6.13)

where c_{14} is a constant.

Thus, for $\zeta = 0$, we may take for the polynomial $P_0(z)$ the polynomial $Q(z) - 4\delta b Q^2(z)$, for the relations (8.6.10) and (8.6.13) are equivalent to (8.6.7) and (8.6.8) respectively, in the statement of the lemma. To complete the proof of the lemma, we remark first that (8.6.10) and (8.6.13) hold, not only for $\zeta = 0$, but also in some neighbourhood $|\zeta| < \epsilon(0)$ of $\zeta = 0$. Since E is closed and since each point $\zeta_0 \in E$ is covered by some circle $|\zeta - \zeta_0| < \epsilon(\zeta_0)$, it follows from the Borel covering theorem that there exists a finite number of points $\zeta, \zeta_2, ..., \zeta_\nu$, such that any point $\zeta \in E$ lies in at least one of the circles $|\zeta - \zeta_i| < \epsilon(\zeta_i)$, where $\zeta_i \in E$, $i = 1, 2, ..., \nu$. Since (8.6.7) and (8.6.8) can be proved for each point $\zeta \in E$, we thus have found a finite set $\zeta_1, ..., \zeta_\nu$ such that (8.6.7) and (8.6.8) hold in $|\zeta - \zeta_i| < \epsilon(\zeta_i)$. For a definite point $\zeta \in E$, suppose that ζ lies inside the circles $|\zeta - \zeta_{k_j}| < \epsilon(\zeta_{k_j})$, $j = 1, ..., m$. Since the k_j's are distinct, we may associate with ζ a unique integer $N = \min\{k_1, ..., k_m\}$, so that the degree of the polynomial $P_\zeta(z)$ is bounded by an integer independent of ζ. This completes the proof of the lemma.

We now define a polynomial $\mathscr{P}(z)$ by the formula

$$\mathscr{P}(z) = \frac{1}{2\pi} \iint_{E-H\delta} \frac{P_\zeta(z)}{\zeta - z} \, d\xi \, d\eta,$$

whence it follows that

$$\left| \mathscr{P}(z) - \frac{1}{2\pi} \iint_{E-H_\delta} \frac{\Psi'(\zeta)}{\zeta - z} \, d\xi \, d\eta \right|$$
$$\leqslant \frac{1}{2\pi} \iint_{E-H_\delta} |\Psi'(\zeta)| \left| \frac{1}{\zeta - z} - P_\zeta(z) \right| d\xi \, d\eta.$$

For that part of G' of $E - H_\delta$ in $|\zeta - z| > 10\delta$, we have, by (8.6.8),

$$\left| P_\zeta(z) - \frac{1}{\zeta - z} \right| < \frac{c_2 \delta^2}{|\zeta - z|^3} < \frac{c_2}{10^3 \delta},$$

whence, by means of (8.6.3), we have

$$\frac{1}{2\pi} \iint_{G'} |\Psi'(\zeta)| \left| \frac{1}{\zeta - z} - P_\zeta(z) \right| d\xi \, d\eta < c_{15} \omega(\delta),$$

where c_{15} is a constant. For that part G'' of $E - H_\delta$ in $|\zeta - z| \leqslant 10\delta$, we have, by (8.6.7) and (8.6.3),

$$\frac{1}{2\pi} \iint_{G''} |\Psi(\zeta)| \left| \frac{1}{\zeta - z} - P_\zeta(z) \right| d\xi \, d\eta$$

$$\leqslant \frac{A}{2\pi\delta} \iint_{G''} |\Psi(\zeta)| \, d\xi \, d\eta + \frac{1}{2\pi} \iint_{G''} \frac{|\Psi(\zeta)|}{|\zeta - z|} d\xi \, d\eta \leqslant c_{16} \, \omega(\delta),$$

where c_{16} is a constant. This proves the theorem.†

Before applying Theorem 8.10 to the theory of cluster sets, we state two important corollaries.

COROLLARY 6.‡ *If E is a bounded continuum of plane measure zero not separating the plane, then any function continuous on E can be approximated on E to any degree of accuracy by means of a polynomial.*

COROLLARY 7.§ *If E is a compact set, then a necessary and sufficient condition that a function, which is continuous on E, can be expanded into a series of polynomials uniformly convergent on E, is that E be a nowhere dense set which does not separate the plane.*

7. An existence theorem on curvilinear cluster sets

We shall use Theorem 8.10 to show that the radial boundary behaviour of an analytic function on radii, or more generally, the behaviour on a family of mutually disjoint arcs in $|z| < 1$, terminating at a set E of first category on $|z| = 1$, can be prescribed with a high degree of freedom. More specifically, we denote by ψ an arbitrary homeomorphism of $|z| \leqslant 1$ onto itself which leaves each point of $|z| = 1$ fixed, by C_n the circle $|z| = 1 - (1/n)$, $n = 2, 3, \ldots$, and by F_n the closure of the finite domain bounded by the Jordan curve $\psi(C_n)$. For any $e^{i\theta}$, let λ_θ denote the radial segment of $|z| \leqslant 1$ terminating at $e^{i\theta}$, and let L_θ denote the simple arc $\psi(\lambda_\theta)$ terminating at $e^{i\theta}$; we observe that L_θ may be tangential

† An alternative proof of Theorem 8.10 has recently been given by Carleson [2]. See also Browder [1].

‡ Hartogs and Rosenthal [1].

§ Lavrentiev [1]. A simple, direct proof of Corollary 7 has been given by Mergelyan [2].

and that points of L_θ need not tend monotonely to $|z| = 1$. With these preliminaries, we may prove the principal result† of this paragraph.

THEOREM 8.11. *Let $\varphi(z)$ be an arbitrary continuous function in $|z| < 1$, let E be a set of first category on $|z| = 1$, and let $\{L_\theta\}$ be a family of mutually disjoint arcs terminating at the points $e^{i\theta} \in E$ and such that $\{L_\theta\}$ is homeomorphic to a family of radial segments $\{\lambda_\theta\}$ terminating at E. Then there exists a function $f(z)$, analytic in $|z| < 1$, such that for every $e^{i\theta} \in E$,*

$$f(z) - \varphi(z) \to 0 \quad as \quad z \to e^{i\theta} \quad on \quad L_\theta. \qquad (8.7.1)$$

We remark that no restrictions concerning the measure of E have been made, and that (8.7.1) implies that

$$C_{L_\theta}(f, e^{i\theta}) = C_{L_\theta}(\varphi, e^{i\theta}).$$

Since E is of first category, we may write $E = \bigcup_{n=1}^{\infty} e_n$, where e_n is nowhere dense. Let E^* denote the set $\bigcup_{n=1}^{\infty} E_n$, where E_n is the closure of e_n. We shall construct an analytic function $f(z)$ such that (8.7.1) is satisfied for every $e^{i\theta} \in E^*$ and uniformly on each E_n.

We may assume, without loss of generality, that $\varphi(z) = 0$ for $z \in F_2$; we set $\varphi_1(z) \equiv 0$, and let $\varphi_n(z)$ be a continuous function in $|z| \leqslant 1$ such that

$$\varphi_n(z) = \varphi(z) - [\varphi_1(z) + \ldots + \varphi_{n-1}(z)] \quad (z \in F_{n+1}).$$

Then

$$\varphi_n(z) = 0 \quad (z \in F_n) \qquad (8.7.2)$$

and

$$\varphi(z) = \sum_{n=1}^{\infty} \varphi_n(z) \quad (|z| < 1), \qquad (8.7.3)$$

the convergence of (8.7.3) being trivial, since, for a given z, only a finite number of terms are different from zero.

We denote by K_n the compact set consisting of the union of F_n and those arcs L_θ terminating at the points of E_n. By (8.7.2), $\varphi_n(z)$ is analytic at every point belonging to the interior of K_n, so that, by Theorem 8.10, there exists a sequence of polynomials $p_n(z)$ with the property that

$$|p_n(z) - \varphi_n(z)| < 2^{-n} \quad (z \in K_n, \, n \geqslant 1). \qquad (8.7.4)$$

† Bagemihl and Seidel [2], Rudin [1]; the proof given here is due to Rudin.

If we set $\qquad g(z) = \sum\limits_{n=1}^{\infty} p_n(z) \quad (|z| < 1),$ $\qquad\qquad$ (8.7.5)

it follows from (8.7.2) and (8.7.4) that $|p_n(z)| < 2^{-n}$ on F_n, whence (8.7.5) converges uniformly on every compact subset of $|z| < 1$, so that $g(z)$ is analytic in $|z| < 1$.

Since the finite sum in the expression

$$g(z) - \varphi(z) = \sum_{n=1}^{N-1} \{p_n(z) - \varphi_n(z)\} + \sum_{N}^{\infty} \{p_n(z) - \varphi_n(z)\}$$

is continuous in $|z| \leqslant 1$, it follows from (8.7.4) that the series

$$\sum_{N}^{\infty} \{p_n(z) - \varphi_n(z)\}$$

is continuous on K_n. Thus there exists a function $\psi(\theta)$, defined on E^* and continuous on each E_n, such that

$$\lim_{z \to e^{i\theta}} \{g(z) - \varphi(z)\} \to \psi(\theta) \quad (z \in L(\theta)),$$ \qquad (8.7.6)

for each $e^{i\theta} \in E^*$.

The theorem will be proved by constructing a function $h(z)$, analytic in $|z| < 1$ and continuous on each K_n, such that

$$h(e^{i\theta}) = \psi(\theta) \quad (e^{i\theta} \in E^*),$$ $\qquad\qquad$ (8.7.7)

for the function $f(z) = g(z) - h(z)$ will then satisfy (8.7.1). With the help of Theorem 8.10 we form by induction a sequence of polynomials $q_n(z)$: having chosen $q_1(z), \ldots, q_n(z)$, $n \geqslant 1$, such that

$$|\psi(\theta) - q_1(e^{i\theta}) - \ldots - q_n(e^{i\theta})| < 2^{-n} \quad (e^{i\theta} \in E_n),$$ \qquad (8.7.8)

we define a function $\psi_n(z)$, continuous on $E_{n+1} \cup K_n$, such that

$$\left. \begin{aligned} \psi_n(z) &= \psi(\theta) - q_1(e^{i\theta}) - \ldots - q_n(e^{i\theta}) && (e^{i\theta} \in E_{n+1}) \\ \psi_n(z) &= 0 && (z \in F_n) \\ |\psi_n(z)| &< 2^{-n} && (z \in K_n) \end{aligned} \right\}$$ \quad (8.7.9)

and approximate $\psi_n(z)$ on $E_{n+1} \cup K_n$ in such a way that

$$|q_{n+1}(z)| < 2^{-n} \quad (z \in K_n)$$

and $\qquad |\psi(z) - q_1(z) - \ldots - q_{n+1}(z)| < 2^{-n-1} \quad (z \in E_{n+1}).$

If we set $h(z) = \sum_{n=1}^{\infty} q_n(z)$, it follows that this series converges uniformly on each K_n, and (8.7.8) implies (8.7.7). This completes the proof of Theorem 8.11.

8. Some consequences of the existence theorem

Because of the ease with which a continuous function with pathological boundary behaviour may be constructed in $|z| < 1$, it is clear that Theorem 8.11 shows the existence of an analytic function in $|z| < 1$ having the same boundary behaviour on a set of first category. The most interesting case, in view of uniqueness theorems such as Theorem 8.1 of this chapter, is that in which the set E of first category is of positive measure and, in particular, of measure 2π on $|z| = 1$. In the corollaries which follow, the set E is presumed to be of first category on $|z| = 1$, and we remark that the measure $m(E)$, $0 \leqslant m(E) \leqslant 2\pi$, may be arbitrary.

COROLLARY 8. *For any continuum \mathscr{K} on the Riemann sphere, there exists a function $f(z)$, analytic in $|z| < 1$, such that*

$$C_\rho(f, e^{i\theta}) = \mathscr{K}$$

for each $e^{i\theta} \in E$.

For each $n = 1, 2, \ldots$, we cover \mathscr{K} with a finite number of circles of (spherical) radius $1/n$ to form a connected set G_n containing \mathscr{K}. Let $0 = r_0 < r_1 < r_2 \ldots < r_n < \ldots$ be a sequence with $\lim_{n \to \infty} r_n = 1$, and let $g(r)$ be a complex-valued continuous function for $0 \leqslant r < 1$ such that the image of the interval $[r_n, r_{n+1}]$ under $g(r)$ is a curve lying in G_n and intersecting each circle of which G_n is formed. If we set $\varphi(z) = g(|z|)$, then $\varphi(z)$ is continuous in $|z| < 1$ and $C_\rho(\varphi, e^{i\theta}) = \mathscr{K}$ for each $e^{i\theta} \in E$. By Theorem 8.11, there exists an analytic function with the same property, and the corollary is proved.

The argument of the proof of Corollary 8 must be modified to treat the case in which the continuum \mathscr{K} reduces to a point, for otherwise it is conceivable that the resulting function $f(z)$ may reduce to a constant. The idea of the modification is contained in the proof of Theorem 8.11; indeed, since E is of first category, we may find a point $e^{i\theta_0}$ not in E and add to the union $\bigcup_{n=1}^{\infty} e_n$ the set e_0

consisting of the single point $e^{i\theta_0}$. If \mathscr{K} consists of a single point, say $w = 0$, we stipulate that $\varphi(z) \to 0$ as $|z| \to 1$ along any radius drawn to a point of E and that $\varphi(z) \to \alpha \neq 0$ as $|z| \to 1$ along the radius $\theta = \theta_0$. Theorem 8.11 now implies the existence of an analytic function $f(z)$ having 0 as a radial limit at every point of E, while the existence of a radius along which $f(z) \to \alpha$ shows that $f(z)$ cannot be constant. Hence we have

COROLLARY 9. *For any complex number w_0 and for any set E of first category on $|z| = 1$, there exists a non-constant function $f(z)$, analytic in $|z| < 1$, with the property that $\lim_{r \to 1} f(r\, e^{i\theta}) = w_0$ for every $e^{i\theta} \in E$.*

We remark that Corollary 9 implies the existence of an analytic function whose real and imaginary parts have prescribed radial limit values on a set E of first category on $|z| = 1$; the measure of E may be 2π. Theorem 8.11 exposes, in a clear way, the dependence of certain radial boundary properties of an analytic function on the notion of category; for further modifications and refinements in this area, the reader is referred to Bagemihl and Seidel [2].

CHAPTER 9

PRIME ENDS

1. Introduction

In Chapter 3 we discussed the correspondence between the
boundary C of a Jordan domain D and the circumference K of
the unit disc U: $\{z:|z| < 1\}$ under a conformal mapping of the
disc onto the simply connected open set D; we also proved
(Theorem 3.2) that the mapping induces a one-to-one correspond-
ence between the individual points of K and the individual points
of $C = \operatorname{Fr} D$ and that this correspondence is in fact topological.
If $w = f(z)$ is the mapping function and $e^{i\theta}$ is an arbitrary point
of K, then the cluster set $C(f, e^{i\theta})$ is degenerate in this case, and
we may write $f(e^{i\theta}) = C(f, e^{i\theta})$, where $f(e^{i\theta})$ is a continuous one-
valued function of $e^{i\theta}$ in $0 \leqslant \theta \leqslant 2\pi$, giving a topological mapping
of K onto $\operatorname{Fr} D$. This can no longer be asserted if the condition
that $\operatorname{Fr} D$ be a Jordan curve is removed and the only condition
imposed on D is that it be a simply connected domain. By the
Riemann mapping theorem† there is a univalent regular function
$w = f(z)$ which maps the open disc U conformally onto D in this
general case also. The cluster set $C(f, e^{i\theta})$ of the mapping function
$f(z)$ then exists for every point $e^{i\theta} \in K$, but we have no assurance
that it is degenerate at any given point, or indeed, at any point
of K. However, $f(z)$ being univalent, the range $R(f, e^{i\theta})$ is empty,
so that, by Theorem 4.2 of Chapter 4, $C(f, e^{i\theta})$ is a frontier set in
the sense that $C(f, e^{i\theta}) = \operatorname{Fr} C(f, e^{i\theta})$. Further, by the mapping
theorem, no point $w \in C(f, e^{i\theta})$ can be an interior point of D; and,
by the definition of $C(f, e^{i\theta})$, no point of $C(f, e^{i\theta})$ can be exterior
to \bar{D}. Therefore $C(f, e^{i\theta}) \subset \operatorname{Fr} D$.

The question now arises whether, in this general case, a con-
formal mapping of the unit disc onto D induces a correspondence
between K and $\operatorname{Fr} D$. The solution of this problem is due, like
that of the Jordan domain, to Carathéodory [1, 2], who proved
that such a conformal mapping induces a one-to-one correspond-

† See, for example, Littlewood [1], p. 135.

ence between the points of K and certain elements of the frontier $\operatorname{Fr} D$ which he defined and called *prime ends*.† More than this, the set \mathscr{E} of prime ends of D gives a compactification $D^* = D \cup \mathscr{E}$ of D; and the conformal mapping of the open disc U onto D induces a homeomorphism between the closed disc \overline{U} and the compactification D^*.

The analogy with the correspondence between a point $e^{i\theta}$ of K and the corresponding point $f(e^{i\theta}) = C(f, e^{i\theta})$ of the boundary C of a Jordan domain suggests a correspondence under conformal mapping in the case of the general simply connected domain D between $e^{i\theta}$ and $C(f, e^{i\theta})$. This turns out to be the case in the sense that $C(f, e^{i\theta})$ coincides with what we shall define as the *impression* of the prime end which corresponds to the point $e^{i\theta}$ of K under the mapping. We must be careful, however, to distinguish between a prime end and its impression since several prime ends, and indeed an infinity of them, may have the same impression.

2. Definition of a prime end

Certain preliminaries are necessary before we can give the definition. We shall denote by D a bounded simply-connected domain in the w-plane without restrictive conditions on its frontier $\operatorname{Fr} D$.

A simple (Jordan) arc with one end-point on $\operatorname{Fr} D$ and all its other points in D is called an *end-cut* of D. A Jordan arc that lies in D except for its two end-points or a Jordan curve (homeomorph of the unit circumference) that lies in D except for one point is a *cross-cut* of D.

A point of $\operatorname{Fr} D$ is *accessible* from D if it is an end-point of an end-cut in D. It is *rectilinearly accessible* if the end-cut is a rectilinear segment.

We shall say that a set of points $E \subset \operatorname{Fr} D$ is dense in $\operatorname{Fr} D$ if

† An equivalent definition, preferred by some writers, was given shortly afterwards by Koebe [1]. We have adopted Carathéodory's definition, since its relation to the cluster set under a mapping is quite transparent. Both definitions however are topological and express intrinsic properties of the domain D independent of any mapping. In its modern context the term *prime end* is perhaps not entirely happy, and Koebe's term *boundary element* (*Randelement*) would be preferable were it not that *prime end* is consecrated by long usage and is best retained for that reason.

every circle having its centre α in Fr D contains a point $\beta \in E$ where $\beta \neq \alpha$. With this definition we have the trivial

THEOREM 9.1. *The points of* Fr D *rectilinearly accessible from* D *are dense in* Fr D.

For let α be any point of Fr D and x any point in $U(\alpha, \epsilon) \cap D$, where $\epsilon > 0$ is such that $U(\alpha, \epsilon)$ does not contain the continuum Fr D. If α is not rectilinearly accessible, the point of Fr D on the segment αx and nearest to x is a rectilinearly accessible point in $U(\alpha, \epsilon)$. If α is rectilinearly accessible, we seek to find another such point in $U(\alpha, \epsilon)$. Let γ be the circumference $|w - \alpha| = \frac{1}{2}\epsilon$ and α' a point of $\gamma \cap$ Fr D. We can find $x' \in \gamma \cap D$ which is not on the diameter of γ through α', since D is an open set. The point of Fr D on the segment $\alpha'x'$, which does not pass through α, and nearest to x' is a rectilinearly accessible point inside or on $\gamma \in U(\alpha, \epsilon)$, so that the theorem is proved.

REMARK. If we drop the condition *rectilinearly* we have already proved more than this. For the corollary to Lemma 3 of Chapter 6 shows that the set of accessible points is of positive linear measure in every neighbourhood of any point of Fr D.

A sequence $q_1, q_2, \ldots, q_n, q_{n+1}, \ldots$ of cross-cuts of D is called a *chain* if the following conditions are satisfied:

 (i) no two of them have any point, including their end-points, in common;

 (ii) q_n separates D into two domains, one of which contains q_{n-1} and the other q_{n+1};

 (iii) the diameter of q_n tends to zero as $n \to \infty$. It follows from (ii) that, of the two subdomains of D determined by q_n, one, denoted by d_n, contains all the cross-cuts q_ν, $\nu > n$, while the other contains all the cross-cuts q_ν, $\nu < n$.

Two chains $Q = \{q_n\}$ and $Q' = \{q'_n\}$ in D are *equivalent* if, for all values of n, the domain d_n contains all but a finite number of the cross-cuts q'_n and the domain d'_n defined by q'_n contains all but a finite number of the cross-cuts q_n. The class of all chains equivalent to a given chain is an *equivalence class*.

A prime end of D is an equivalence class of chains in D. A chain belonging to such a class is said to belong to the prime end, which we shall generally denote by P.

If P is a prime end of D, $\{q_n\}$ is a chain belonging to P, and d_n is the subdomain of D defined by q_n and containing q_{n+1}, and if $\{q_n'\}$ is an equivalent chain, then evidently

$$\bigcap \bar{d}_n \subset \bigcap \bar{d}_n' \quad \text{and} \quad \bigcap \bar{d}_n' \subset \bigcap \bar{d}_n,$$

so that $\bigcap \bar{d}_n = \bigcap \bar{d}_n'$. The set

$$I(P) = \bigcap \bar{d}_n,$$

which is invariant in the equivalence class of chains constituting P, is called the *impression* of the prime end P. Two distinct prime ends of D may have the same impression. For example, if the domain D is obtained by deletion of an end-cut γ from the unit disc, then each interior point of γ constitutes the impression of precisely two prime ends. It follows from § 1 of Chapter 1 that $I(P)$ *is either a continuum or a single point*.

The domains d_n form a nested sequence. A sequence of points p_n or sets E_n of D such that every domain d_n contains all but a finite number of them is said to *converge* to the prime end P.

Since the diameter of q_n tends to zero as $n \to \infty$, the chain $\{q_n\}$ contains an equivalent subchain which tends to a limit $\omega \in \operatorname{Fr} D$. This subchain is said to be a *convergent chain* belonging to or defining P. The limit ω is not necessarily unique; there may in fact be a continuum of such limits of distinct but equivalent convergent chains belonging to P. But it follows from the foregoing remarks that *a prime end may always be defined by a convergent chain*. We may in fact particularize still further and impose the condition that the cross-cuts of the convergent chain shall be concentric circular arcs converging to their common centre. This will be proved in Theorem 9.3. But first we prove

THEOREM 9.2. *For any prime end P of D,*

$$I(P) \subset \operatorname{Fr} D.$$

Let $\{q_n\}$ be a convergent chain belonging to P, let $\{d_n\}$ be a nested sequence of domains defined by the chain, and let $\omega = \lim_{n \to \infty} q_n$. If now O is a point of D outside d_1 and A is any other point of D, these two points can be joined by a polygonal path L in D whose distance from $\omega \in \operatorname{Fr} D$ is positive. Since q_n meets $\operatorname{Fr} D$ and diam $q_n \to 0$ as $n \to \infty$, L meets only a finite number of

the cross-cuts q_n. It follows that A is contained in only a finite number of the domains d_n and hence is not contained in $I(P) = \bigcap \bar{d}_n$. Since A is any point of D, and since an exterior point of D is not contained in the closure \bar{d}_n of any d_n, the theorem is proved.

Figures 9.1–9.3 (on pp. 180 and 181) illustrate the notion of a prime end.

3. Preliminary lemmas

LEMMA 1. *If w_1, w_2 are two points in the simply connected domain D and K_0 is a circle of centre $\omega \in \mathrm{Fr}\, D$ that separates the two points, then some arc of K_0 is a cross-cut of D that separates w_1 from w_2.*

This is a special case of a general topological theorem,† but we give an independent proof.

Since D is an open set, the points w_1, w_2 can be joined by a polygonal arc $\gamma \in D$ consisting of a finite number of segments whose distance δ from $\mathrm{Fr}\, D$ is positive. The number of crossings of γ and K_0 is finite and odd. Also, the intersection $K_0 \cap D$ is a finite or countable set of arcs of K_0 of which, since $\delta > 0$, only a finite number can meet γ. It follows that there is at least one arc $\alpha \in K_0$ which is a cross-cut of D and is crossed by γ in an odd number of points. If the number of crossings of γ and α is greater than one, then, without altering the end-points w_1, w_2, the arc γ can be deformed into a polygonal path γ' which crosses α in just one point β. The arc of γ' joining w_1 to β lies in one component of $D - \alpha$ and that joining w_2 to β in the other, so that α separates w_1 from w_2 in D, and the lemma is proved.

We note expressly that there may be more than one arc of K_0 with the same property as α.

THEOREM 9.3. *A prime end contains a chain of cross-cuts that lie on a sequence of concentric circles whose radii tend to zero.*

The prime end contains a convergent chain $\{q_n\}$, $\operatorname{diam} q_n \to 0$ and $\lim q_n = \omega$, say, as $n \to \infty$, $\omega \in \mathrm{Fr}\, D$. Given $\epsilon > 0$, construct the circle $K(\omega, \epsilon)$: $\{w : |w - \omega| = \epsilon\}$. We may assume that an interior point O of D lies outside $K(\omega, \epsilon)$. All except a finite number of the q_n lie inside $K(\omega, \epsilon)$. By Lemma 1 there is an arc $\alpha(\epsilon)$ of

† See, for example, Newman [1], p. 151, Theorem 7.1.

$K(\omega, \epsilon)$ forming a cross-cut of D which separates O from q_N, say, inside $K(\omega, \epsilon)$ and hence, since $\{q_n\}$ is a chain, from all q_n, $n \geqslant N$. Now choose $\epsilon_1 < \frac{1}{2}\epsilon$ such that q_N is outside $K(\omega, \epsilon_1)$. Again, all except a finite number of the q_n are inside $K(\omega, \epsilon_1)$. There is therefore a cross-cut $\alpha(\epsilon_1)$ of D on $K(\omega, \epsilon_1)$ which separates q_N and therefore also $\alpha(\epsilon)$ from all the q_n for which $n > N_1 > N$, say. Continuing this process we obtain a sequence of circular arcs $\alpha(\epsilon_\nu)$ converging to ω and forming a chain equivalent to the convergent chain $\{q_n\}$ and therefore contained in the prime end. This proves the theorem.

We remark that it follows from this theorem that *a prime end may always be defined by a chain of concentric cross-cuts of the domain of strictly decreasing radii converging to their common centre.*

4. Correspondence of frontiers under conformal mapping

These preliminaries enable us to prove that a conformal mapping $w = f(z)$ of the open unit disc $U = \{z : |z| < 1\}$ onto a simply connected domain D, which we may take to be bounded but not otherwise restricted, induces a one-to-one correspondence between the points of $K = \{z : |z| = 1\}$ and the prime ends of D. The proof is the same in principle as in the simpler case of a Jordan domain (cf. Theorem 3.2).

We take $z = 1$ as a typical point of K. By Theorem 3.1 there exists a sequence of disjoint circular cross-cuts γ_ν of U converging to $z = 1$ whose images $f(\gamma_\nu)$ under a conformal mapping $w = f(z)$ of U onto D form a chain defining a prime end which we denote by $P(1)$. In the notation of Chapter 3, the sub-domain of U cut off by γ_ν and containing $\gamma_{\nu+1}$ is denoted by d_ν and its image $f(d_\nu)$ by \mathscr{D}_ν. Then, by the definitions, we have

$$C(f, 1) = \bigcap_\nu \overline{\mathscr{D}}_\nu = I(P(1)).$$

If now $\{z_n\}$ is any sequence of points converging to $z = 1$, all but a finite number of them, for any ν, are contained in d_ν, and consequently all but a finite number of the image points $w_n = f(z_n)$ are contained in \mathscr{D}_ν, so that, by the definition given in § 2, the sequence $\{w_n\}$ corresponding to $\{z_n\}$ converges to $P(1)$,

while the closure of the sequence $\{w_n\}$ is contained in $P(1)$. It follows that *as z approaches an arbitrary point $e^{i\theta} \in K$ on any set of points in U, its image $f(z)$ tends to a unique prime end of D which may be denoted by $P(e^{i\theta})$,* or, if we wish to specify the mapping function, by $P(f, e^{i\theta})$. *Further, $C(f, e^{i\theta}) = I(P(e^{i\theta}))$.*

We have now to prove the converse, namely, that given a prime end P of D, every sequence $\{w_n\} \subset D$ which converges to P has for its pre-image under the conformal mapping $w = f(z)$ a sequence $\{z_n\}$, $|z_n| < 1$, which converges to a unique point $e^{i\theta} \in K$.

The sequence $\{z_n\}$ of the pre-images of the sequence $\{w_n\}$ converging to P has no limit point in U and contains a convergent subsequence $\{z_m\}$, $\lim_{m \to \infty} z_m = e^{i\theta}$, say. Again by Theorem 3.1, there exists a sequence of cross-cuts γ_μ of U converging to $e^{i\theta}$ whose images $f(\gamma_\mu)$ form a chain defining a prime end $P(e^{i\theta})$. The sequence $\{w_m\}$, $w_m = f(z_m)$, converges to $P(e^{i\theta})$ and, since it also converges to P, we have $P(e^{i\theta}) = P$. If $\{z_n\}$ contains another subsequence $\{z_p\}$ converging to a point $e^{i\varphi}$, where $\varphi \neq \theta$, there is a sequence of cross-cuts γ_ν converging to $e^{i\varphi}$ whose images $f(\gamma_\nu)$ form a chain defining a prime end $P(e^{i\varphi})$. But, for sufficiently large μ, ν the domains δ_μ defined by γ_μ do not meet any of the domains δ_ν defined by γ_ν, so that the image domains d_μ do not meet the image domains d_ν. Since every δ_ν contains all but a finite number of the points z_p, every d_ν contains all but a finite number of the image points w_p. By hypothesis, $\{w_p\}$ is a subsequence of the sequence $\{w_n\}$ which converges to $P(e^{i\theta})$, so that all but a finite number of the points w_p are contained in d_μ which is disjoint from d_ν. We thus have a contradiction from which it follows that $\varphi = \theta$. This completes the proof of Carathéodory's correspondence theorem, namely,

THEOREM 9.4. *Under a conformal mapping $w = f(z)$ of the open disc U onto a simply connected domain D, there exists a one-to-one correspondence between the points $e^{i\theta} \in K$ and the prime ends $P(e^{i\theta})$ of D. Moreover, $C(f, e^{i\theta}) = I(P(e^{i\theta}))$.*

5. Elementary properties of the space of prime ends

By the introduction of a suitable metric the aggregate of prime ends of D becomes a space which we denote in general by \mathscr{E} or, if

it is necessary to specify the domain, by $\mathscr{E}(D)$. Given any conformal mapping of U onto D, we may adopt as the metric on \mathscr{E} the euclidean distance between the pair of points of K corresponding to a given pair of prime ends (elements) of \mathscr{E}. Under this metric we have

THEOREM 9.5. \mathscr{E} *is a complete metric space.*

For, by Theorem 9.4, a fundamental sequence in \mathscr{E} is the image of a fundamental sequence in K whose limit exists and corresponds to an element of \mathscr{E} which, in this metric, is the limit of a fundamental sequence. \mathscr{E} is thus a complete metric space and hence, by a well-known theorem,† is of second category.

As we shall see later, this theorem is important because it makes Baire's category theorem available for the investigation of the space of prime ends.

We call a subdomain of D a domain d which is defined by a circular cross-cut of D and does not contain some fixed interior point of D. A subset \mathscr{S} of \mathscr{E} is said to be nowhere dense if every subdomain d contains another subdomain d_1 which has no prime end belonging to \mathscr{S}. The prime ends to which the circular cross-cut defining d_1 converges at its end-points are assumed not to belong to d_1. Evidently a nowhere dense subset of \mathscr{E} corresponds under a conformal mapping to a nowhere dense set on K. A subset of \mathscr{E} is of category I if it is the union of a countable set of nowhere dense subsets of \mathscr{E}. *A set $\mathscr{H} \subset \mathscr{E}$ of category I therefore corresponds to a set h of category I on K*, the complements with respect to \mathscr{E} and K being residual.

The prime ends $P \in \mathscr{E}$ have a cyclic order corresponding to the order of the points of K which are their pre-images induced by the mapping, and any pair (P_1, P_2) of them defines an interval of prime ends, i.e. the set of prime ends $P \in \mathscr{E}$ which follow P_1 and precede P_2 in this cyclic order. The mapping by $f(z)$ of U onto D thus induces a Hausdorff topology on the space \mathscr{E}; that is to say, distinct elements of \mathscr{E} are contained in disjoint intervals of \mathscr{E}. Further, *the induced mapping of K onto \mathscr{E}, being a one-to-one continuous mapping of the compact space K onto the compact space \mathscr{E}, is a homeomorphism.*‡

† See Banach [1], p. 14. ‡ Newman [1], p. 70, Theorem 9.3.

We may extend the ordinary topology in D to $D^* = D \cup \mathscr{E}$ by defining neighbourhoods as follows. If d is a subdomain of D as defined above, the set $d \cup \mathscr{E}_d$, where \mathscr{E}_d consists of those prime ends of D that contain a chain whose elements all lie in D (except their end-points), shall be a neighbourhood of each point in d and each prime end in \mathscr{E}_d. Carathéodory's theorem (Theorem 9.4) may accordingly be stated in the following form:

THEOREM 9.6. *A conformal homeomorphism between the open unit disc U and a bounded simply connected domain D can be extended to a homeomorphism between \overline{U} and the prime end compactification D^* of D.*

6. Alternative metrics and generalizations

The procedure of §5 depends essentially on applying the euclidean metric of the closed disc \overline{U} to D^* by means of a particular conformal mapping $f(z)$. Since the Riemann mapping theorem is quite general for simply connected domains, this is always possible and is sufficient for our immediate purposes. But it is natural to seek to base the theory on the intrinsic properties of D without the intervention of a particular mapping. We require a metric in terms of which a fundamental sequence in D converges either to a point of D or to a prime end $P \in \mathscr{E}$ and which is compatible with the topology characterized by the neighbourhoods $d \cup \mathscr{E}_d$ as defined in §5. Here it must suffice to say that a metric based upon the notion of relative distance in D, introduced by Mazurkiewicz [1] and called by him the prime-end metric, achieves this objective. Alternative theories for the plane simply connected domain to achieve the same objective have also been developed by Choquet [1–3], Lelong-Ferrand [1–3], Cesari [1], and Schlesinger [1].

A modified definition of prime ends equivalent to that of Carathéodory, as far as mapping theory is concerned, has been given by Collingwood and Piranian [4]. This is based on the notion of a 'weak chain' for which the condition (i) of the definition of a chain is replaced by the condition rel. dist. $(q_n, q_{n+k}) > 0$. The advantage of this extension of the definition is that the proof of Theorem 9.4 no longer depends on the deep theorem of F. and M.

Riesz (Theorem 2.5), so that a weaker form of Tsuji's theorem (Theorem 3.1), in which the cross-cuts of D are not required to have both end-points distinct from those of all the rest, suffices. Since, however, the Riesz theorem is required for other purposes we have retained the classical definition of Carathéodory.

The theory of prime ends is easily extended to plane domains of finite connectivity; but for more general domains in the plane or for open sets in spaces of higher dimensions, the problem, which has been studied by Kaufmann [1], Mazurkiewicz [2], Freudenthal [1], and others, becomes decidedly more formidable and lies outside the scope of this book. A theory for variable domains has also been developed by Suvorov [1].

7. Principal and subsidiary points

A point $p \in I(P)$ is a *principal point* relative to the prime end P of D if every neighbourhood $U(p, \epsilon)$ contains a cross-cut q of D belonging to a chain $\{q_n\}$ belonging to P. Thus, p is the limit of a convergent chain belonging to P. Since every chain belonging to P has at least one limit point in $I(P)$, *the set $\Pi(P)$ of principal points of $I(P)$ is not empty and is closed.* A point of $I(P)$ which is not a principal point is a *subsidiary point* relative to P.

Now let $L = L(t)$, $0 \leqslant t < 1$, be a continuous curve in D converging to P, i.e. given any chain $\{q_n\}$ belonging to P and any large N, there exists $t_N < 1$ such that $L(t), t_N < t < 1$, is contained in d_N, where $\{d_n\}$ is the nested sequence of domains defined by $\{q_n\}$. Given any $\omega \in \Pi(P)$, we can find a chain $\{q_n(\omega)\}$ converging to ω, so that, because L intersects q_n for all large n, the cluster set $C(L, 1)$ of the curve L contains ω. Therefore

$$\Pi(P) \subset C(L, 1). \tag{9.7.1}$$

Since (9.7.1) holds for any $L = L(t)$ converging to P (which we denote by $L \operatorname{cgt} P$) and since, by definition of convergence, $C(L, 1) \subset I(P)$, we have

$$\Pi(P) \subset \bigcap_{L \operatorname{cgt} P} C(L, 1) \subset I(P), \tag{9.7.2}$$

where the intersection is taken over all Jordan curves L in D converging to P. We now prove

THEOREM 9.7. *A prime end P whose impression $I(P)$ contains an accessible point relative to P has only one principal point. Moreover, there can be at most one accessible point relative to P which, if it exists, is the sole principal point.*

Suppose that $\Pi(P)$ contains two distinct (principal) points a and b. There then exist two equivalent chains $\{q_n(a)\}$, $\{q_n(b)\}$ belonging to P and converging respectively to a and b. Any Jordan curve $L(t)$, $0 \leqslant t < 1$, in D and converging to P intersects all except a finite number of cross-cuts in both these chains, so that a and b both belong to $C(L, 1)$. Therefore L cannot be an end-cut of D and consequently $I(P)$ contains no accessible point relative to P.

To prove the second part, observe first that *every end-cut of D converges to a prime end*. For suppose that

$$\gamma(t), \quad 0 \leqslant t < 1, \quad \lim_{t \to 1} \gamma(t) = \alpha = C(\gamma, 1),$$

where $\alpha \in \mathrm{Fr}\, D$, is the end-cut and that $\{t_n(1)\}$, $\{t_n(2)\}$ are two sequences such that

$$t_n(1) < t_n(2) < t_{n+1}(1) \quad \text{and} \quad \lim_{n \to \infty} t_n(1) = \lim_{n \to \infty} t_n(2) = 1,$$

while the corresponding sequences $\{\gamma(t_n(1))\}$ and $\{\gamma(t_n(2))\}$ converge to distinct prime ends P_1 and P_2. Denoting the points $\gamma(t_n(1))$ and $\gamma(t_n(2))$ of γ by $p_n(1)$ and $p_n(2)$, it follows from Theorem 9.3 that for all $n > N$, say, the arc of γ joining $p_n(1)$ to $p_n(2)$ intersects a fixed cross-cut q of D which we may assume to be at a positive distance from α. Taken over all values of $n > N$ these intersections have at least one limit point $\beta \in \bar{q} \neq \alpha$. But $\beta \in C(\gamma, 1) = \alpha$ and so we have a contradiction.

Now let $\omega \in I(P)$ be an accessible point of $\mathrm{Fr}\, D$ relative to P, so that there is an end-cut $L(t)$ cgt P having $\omega = \lim_{t \to 1} L(t) = C(L, 1)$ as its end-point. Suppose that $\omega' = \Pi(P)$ is the sole principal point. There is then a chain $\{q_n\}$ converging to ω' and belonging to P. Since $L(t)$ cgt P, all except a finite number of the q_n intersect $L(t)$, $1 - \epsilon < t < 1$, where $\epsilon > 0$ is arbitrarily small. Hence

$\omega' \in C(L, 1)$. But $C(L, 1) = \omega$ and hence $\omega = \omega'$. Let us call a continuous arc L cgt P a *principal path* of the prime end P if $C(L, 1) = \Pi(P)$. We have just seen that if L is an end-cut, it is a principal path, but we have not yet proved the existence of a principal path in the case of a prime end having no accessible point. This, however, is implied by the following theorem due to Lindelöf.†

THEOREM 9.8. *Under a conformal mapping $w = f(z)$ of the disc U onto a bounded domain D, the cluster set $C_S(f, e^{i\theta})$ on any Stolz path S in U terminating at $z = e^{i\theta}$ is the set $\Pi(P(e^{i\theta}))$ of principal points of the prime end $P(e^{i\theta})$ corresponding to $e^{i\theta}$ under the mapping.*

The idea of our proof derives from that of Theorem 3.1. We may put $\theta = 0$ and write $z = 1 + r e^{i\varphi}$. By Lemma 1 of Chapter 3 we may construct a sequence of circular cross-cuts γ_n of U, of centre $z = 1$ and radius $r_n \to 0$ as $n \to \infty$, such that $L_n = o(1)$, where

$$L_n = \int_{\gamma_n} |f'(z)| \, |dz|$$

is the length of the image $f(\gamma_n)$ of γ_n.

Next we choose constants α and β, where $0 < \alpha < \beta < \pi$, denote by V_n the 'rectangle'

$$r_{n+1} < r < r_n, \quad 0 < \varphi < \alpha,$$

and write $\qquad \Lambda(n, \varphi) = \int_{r_{n+1}}^{r_n} |f'(r e^{i\varphi})| \, dr.$

Then

$$\int_0^\alpha \Lambda(n, \varphi) \, d\varphi = \int_0^\alpha \int_{r_{n+1}}^{r_n} |f'(z)| \, dr \, d\varphi \leqslant \frac{1}{r_{n+1}} \iint_{V_n} |f'(z)| \, d\sigma$$

$$\leqslant \frac{1}{r_{n+1}} \left(\iint_{V_n} |f'(z)|^2 \, d\sigma \iint_{V_n} d\sigma \right)^{\frac{1}{2}}.$$

Here the first integral on the right tends to zero as $n \to \infty$. The second integral is less than πr_n^2, and since we can choose $\{r_n\}$ so

† Lindelöf's original proof is rather long. The proof given here is a modification (Collingwood and Piranian [4]) of the much simpler proof given by Tsuji [1]. An alternative proof is by means of Doob's one-sided cluster-set theorem (Theorem 5.18); see Doob [13].

that $r_n/r_{n+1} < 4$, there exist values $\varphi_n(\alpha)$ in $0 < \varphi < \alpha$ such that $\Lambda(n, \varphi_n(\alpha)) = o(1)$. Similarly, there exist values $\varphi_n(\beta)$ in $\beta < \varphi < \pi$ such that $\Lambda(n, \varphi_n(\beta)) = o(1)$.

Now let H_n denote the figure obtained by connecting the cross-cuts γ_n, γ_{n+1} of U by the two segments

$$z = 1 + r e^{i\varphi_n(\alpha)}, \quad z = 1 + r e^{i\varphi_n(\beta)},$$

where $r_{n+1} \leqslant r \leqslant r_n$. The preceding argument shows that

$$\operatorname{diam} f(H_n) = o(1).$$

Also $\qquad \operatorname{diam} f(I_n) < \operatorname{diam} f(H_n) = o(1),$ $\qquad\qquad$ (9.7.3)

where I_n denotes the 'trapezoidal' domain defined by the recti-linear segments in H_n and the arcs of γ_n and γ_{n+1} which they intercept and which do not meet the circumference K of U. Given any Stolz path S terminating at $z = 1$, we can construct the figures H_n so that S and the radius to $z = 1$ are both, from a certain point onwards, contained in $\bigcup_n I_n$. It now follows from (9.7.3) that $C_S(f, 1) = C_\rho(f, 1)$, and since any path T in U terminating at $z = 1$ meets all but a finite number of the H_n,

$$C_T(f, 1) \subset C_\rho(f, 1).$$

We conclude that $C_\rho(f, 1) = C_S(f, 1) = \Pi(P, 1)$, which proves the theorem.

The set $\Pi(P)$ is thus either a single point, in which case it is the sole accessible point of FR D relative to P, or it is a continuum. In either case $\Pi(P)$ *is the minimum cluster set of all paths L cgt P.*

COROLLARY 1. *Under the conformal mapping $w = f(z)$ of the unit disc onto D we have*

$$C_{\mathscr{A}}(f, e^{i\theta}) = \Pi(P), \qquad\qquad (9.7.4)$$

where $C_{\mathscr{A}}(f, e^{i\theta})$ is the outer angular cluster set at $e^{i\theta}$.

For if Δ is any Stolz angle at $e^{i\theta}$, it follows from Theorem 9.8 that $C_\Delta(f, e^{i\theta}) \subset \Pi(P)$ and from the relation $C_{\rho(\phi)}(f, e^{i\theta}) = \Pi(P)$, for an appropriate value of φ, that $C_\Delta(f, e^{i\theta}) \supset \Pi(P)$.

8. The classification of prime ends

The prime ends of a simply connected domain D fall into four classes defined as follows:

A prime end P is

(i) *of the first kind* if its impression $I(P)$ is degenerate, i.e. consists of a single point a; every curve L cgt P is then an end-cut to the accessible point $a = I(P) = \Pi(P)$;

Fig. 9.1.

(ii) *of the second kind* if $I(P)$ contains a single accessible point relative to P and some subsidiary points (since $I(P)$ is a continuum the subsidiary points form one or two infinite sets);

(iii) *of the third kind* if $I(P)$ contains no accessible point relative to P and no subsidiary point so that it consists of a continuum $\Pi(P)$ of principal points;

(iv) *of the fourth kind* if $I(P)$ contains a continuum of principal points and some subsidiary points which again form one or more infinite sets.

We have already agreed to denote the set of prime ends of D by \mathscr{E}. The subsets of \mathscr{E} of prime ends of the four kinds will be

Fig. 9.2.

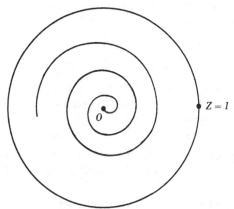

Fig. 9.3.

denoted by \mathscr{E}_ν, $\nu = 1, 2, 3, 4$, and the corresponding sets of points of K under a conformal mapping by e_ν, $\nu = 1, 2, 3, 4$. We have seen in §6 how \mathscr{E} is made a complete metric space by applying the metric of the corresponding space K under the mapping.

The four kinds of prime ends are illustrated by figures 9.1–9.3.

In figure 9.1, DE is the impression of a prime end P_1 of the *second kind* having E as accessible point; $AB = I(P_2)$ and $GH = \Pi(P_2)$ where P_2 is of the *fourth kind*; every point of the boundary of the domain suggested by the figure except the closed segments DE and AB is the impression of a prime end of the *first kind*. In figure 9.2, $CD = \Pi(P_1)$ and $EF = \Pi(P_2)$ are the impressions of prime ends of the *third kind* and all other boundary points are of the *first kind*. In figure 9.3 the unit circumference is the impression of a prime end of the *third kind*.

9. The distribution of the subsets \mathscr{E}_ν in \mathscr{E}

Although considerable progress has been made in recent years, the problem of the distribution in the sense of relative abundance of the four kinds of prime ends of an arbitrary domain is not completely solved. A number of results can, however, be deduced without difficulty from theorems we have already proved.

The most familiar prime end is that of the first kind. All the prime ends of a Jordan domain are of this kind. We begin by proving a general property of the set \mathscr{E}_1.

THEOREM 9.9. *The set \mathscr{E}_1 is of type G_δ.*

To prove this we require the following

LEMMA. *Given a number $h > 0$ the set $\mathscr{D}_h \subset \mathscr{E}$ of prime ends P of D such that the diameter of $I(P)$ is not less than h is closed.*

Suppose that every interval of \mathscr{E} which contains P contains a member of \mathscr{D}_h. Then if $\{D_m\}$ is a nested sequence of subdomains of D converging to P, every set \bar{D}_m contains a subset of diameter not less than h, and therefore $I(P)$ is of diameter not less than h, so that $P \in \mathscr{D}_h$.

Theorem 9.9 now follows since evidently

$$\mathscr{E}_1 = \bigcap_{n=1}^{\infty} \mathscr{C}\mathscr{D}_{1/n}$$

and $\mathscr{C}\mathscr{D}_{1/n} = \mathscr{E} - \mathscr{D}_{1/n}$ is open.

Theorem 9.1 is equivalent to the statement that the accessible (and therefore principal) points of $\mathscr{E}_1 \cup \mathscr{E}_2$ are dense in FR D. The following supplementary theorem is an immediate consequence of Fatou's theorem and its corollaries.

THEOREM 9.10. *In every interval of \mathscr{E} the subset $\mathscr{E}_1 \cup \mathscr{E}_2$ has the cardinal of the continuum and is of type $F_{\sigma\delta}$.*

For, under a conformal mapping of the unit disc onto D, the set $e_1 \cup e_2 \subset K$ has the cardinal of the continuum. Further, by Theorem 2.6, $e_3 \cup e_4$ is of type $G_{\delta\sigma}$, so that the complement $e_1 \cup e_2$ of this set with respect to K is of type $F_{\sigma\delta}$. Since these properties are preserved by the induced homeomorphism between K and \mathscr{E}, the theorem is proved.

COROLLARY 2. $\mathscr{E}_3 \cup \mathscr{E}_4$ *is of type $G_{\delta\sigma}$.*

THEOREM 9.11. *If \mathscr{E}_2 is empty, then*

(i) *\mathscr{E}_1 is residual in \mathscr{E} and the corresponding set $e_1(f)$ under the conformal mapping f is both residual and of measure 2π on K;*

(ii) *$\mathscr{E}_3 \cup \mathscr{E}_4$ is of category I in \mathscr{E} and $e_3(f) \cup e_4(f)$ is of category I and measure zero on K.*

It is enough to prove (i) which implies (ii). Since \mathscr{E}_2 is empty, \mathscr{E}_1 is dense in the complete metric space \mathscr{E} by Theorem 9.10 and the result now follows by Theorem 9.9.†

Theorem 9.11 brings out the relative density of \mathscr{E}_1 whenever the set of prime ends having subsidiary points as well as an accessible point is empty. Our next theorem shows the relative density, for a general domain, of the set $\mathscr{E}_1 \cup \mathscr{E}_3$ of the prime ends without subsidiary points.

THEOREM 9.12. *For any simply connected domain D the union $\mathscr{E}_1 \cup \mathscr{E}_3$ of the prime ends of the first and third kinds is a residual subset of the set \mathscr{E}.*

Let $f(z)$ be a conformal mapping of $|z| < 1$ onto D. Then, by Theorem 9.8, $C_\rho(f, e^{i\theta}) = \Pi(P)$ for all θ and, by Theorem 4.8, $C_\rho(f, e^{i\theta}) = C(f, e^{i\theta}) = I(P)$ on a residual subset of points $e^{i\theta} \in K$. This proves the theorem. We remark that it can be shown that $\mathscr{E}_1 \cup \mathscr{E}_3$ is also a set of type G_δ, but the proof‡ is rather long and will be omitted.

For a Jordan domain, $\mathscr{E} = \mathscr{E}_1$; and for a general domain, by

† See, for example, Munroe [1], p. 69.
‡ Piranian [1], p. 84.

Theorem 9.10, $\mathscr{E}_1 \cup \mathscr{E}_2$ has the cardinal of the continuum. We now have

COROLLARY 3.† $\mathscr{E}_2 \neq \mathscr{E}$ *for any simply connected domain.*

For \mathscr{E}_2 is of category I and \mathscr{E} is of category II.

It has been known from the inception of the theory of prime ends that domains exist for which the set \mathscr{E}_1 of prime ends of the first kind is empty.‡ It follows from Theorem 9.10 that the set \mathscr{E}_2 is locally of the power of the continuum even though it is of category I. The following construction gives an example. We start with the domain D_1 of figure 9.4 which lies in the unit disc and spirals towards the unit circle. We divide its boundary

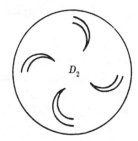

Fig. 9.4. Fig. 9.5.

(exclusive of the unit circle) into arcs of diameter less than $\frac{1}{2}$, and we replace each of these arcs by a pair of curves which lie outside D_1 and spiral towards the unit circle. Clearly this can be done in such a way that no two of the family of new curves intersect except for the common initial point that each has with just one of the others. The domain D_1 has thus been enlarged to a domain D_2 (see figure 9.5). Each of the new curves is now divided into arcs of diameter less than $\frac{1}{4}$, and each of these arcs is again replaced by a pair of curves which spiral towards the unit circle thus enlarging D_2 to a domain D_3. Continuing this process

† This result was first proved as an independent theorem by Weniaminoff [1] and, independently of him, by Urysohn [1], thus answering in the negative the question asked by Carathéodory [1] whether a domain exists for which $\mathscr{E} = \mathscr{E}_2$. For a more general result, see Lohwater and Piranian [1], Theorem 2.

‡ Carathéodory [2], p. 369.

indefinitely we obtain a domain $D_\infty = \bigcup_{k=1}^{\infty} D_k$ which has the desired property.

Let $\{c_m\}$ be any chain of cross-cuts of D_∞. If $\{c_m\}$ converges to a point p inside the unit disc, then the corresponding nested sequence $\{d_m\}$ has the property that each domain d_m contains a spiral domain whose boundary contains the unit circle; in this case p is the principal point of a prime end of the second kind. If $\{c_m\}$ does not converge to a point in the unit disc, then it has no limit point in the disc. In this case it is easily verified that the prime end to which $\{c_m\}$ belongs is of the third kind and that $I(P)$ is the unit circle. Thus, for D_∞, all the (residual) set of prime ends belonging to \mathscr{E}_3 have the same impression so that every point of the unit circumference is, although inaccessible, of infinite multiplicity in the sense that it belongs to an infinity of prime ends.

10. A domain having a bi-connected set of accessible points

We give an example, due to Carathéodory ([2], pp. 369–70) of a domain with the property that $\mathscr{E}_3 \cup \mathscr{E}_4$ is empty while all the prime ends in \mathscr{E}_1 have the same point as impression. The construction is in two stages.

We begin by constructing the triangle ABM in the rectangle $ABCD$, M being the mid-point of CD in figure 9.6. We remove the middle third of each of the sloping sides and join the free ends by straight lines to the intersections with CD of the ordinates through the mid-points of the discarded thirds. We reflect the resulting figure in AB to obtain a polygon \mathscr{P}_2. We then operate in the same way on each side of \mathscr{P}_2 above AB, again reflect in AB and obtain \mathscr{P}_3. Proceeding in this way, by discarding the middle third of every side of the upper half of \mathscr{P}_n and joining the free ends to the intersection with CD of the ordinate through its mid-point and reflecting in AB we obtain \mathscr{P}_{n+1}. Removing the line AB, we have thus obtained a sequence of domains

$$B_1 \subset B_2 \subset \ldots \subset B_n \subset \ldots$$

whose boundaries are the polygons $\mathscr{P}_1, \mathscr{P}_2, ..., \mathscr{P}_n,$ The domain

$$B_\infty = \bigcup_{n=1}^{\infty} B_n$$

has prime ends of the first and second kinds only. It is easily seen that all points of $\text{Fr}\,B_\infty$ on CD and on $C'D'$, its reflexion in AB, are accessible along the ordinates through them and belong to \mathscr{E}_1 and that all other prime ends whose impressions do not meet AB

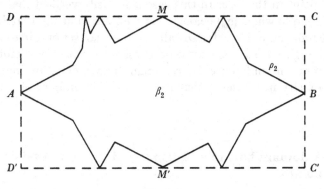

Fig. 9.6.

belong to \mathscr{E}_2.† The points A and B are the accessible points of prime ends of the second kind whose impressions are respectively the segments DD' and CC'. For every other prime end in \mathscr{E}_2 the impression consists of the segment of the ordinate through the accessible point and the nearest of the two lines CD, $C'D'$.

Now let AB, $A = 0$, be the x-axis and AD, $A = 0$, be the y-axis, $z = x + iy$; we make the transformation

$$\xi^2 + \eta^2 = 2(1+x)\xi = 2\left(\frac{1-y^2}{y}\right)\eta; \quad \zeta = \xi + i\eta.$$

This maps the segment DD' onto the circle of unit radius with centre $\zeta = 1$ and maps the segment CC' onto another circle containing the first circle and tangent to it at the origin. CD and $C'D'$ are both transformed into the origin $\zeta = 0$ while AB is

† For the details of the proof, we refer the reader to Carathéodory [2], pp. 366–7.

beginbeginusmessageusmessageususmessageusususususmessageusususmessageusususususmessageusmessageusususmessageususmessageusususmessageusususususususususmessageususususmessageususmessageususmessageususmessageusmessageususmessageususmessageusmessageususmessageusmessageususmessageususususI apologize—my response got corrupted. Let me provide the transcription.

mapped onto the segment $\alpha\beta$ of the ζ-axis intercepted by the crescent-shaped region between the two circles which is the image under the mapping of the rectangle $CDD'C'$ (see figure 9.7). The ordinates $x =$ constant in this rectangle are mapped onto a family of circles like the dotted circle in the figure contained in the crescent and all passing through the origin where they are mutually tangent. B_∞ is mapped onto a domain \mathscr{G} in the crescent. Evidently, there is a one-to-one correspondence under the mapping b between the sets $\mathscr{E}_1(B_\infty)$ and $\mathscr{E}_1(\mathscr{G})$ and between $\mathscr{E}_2(B_\infty)$

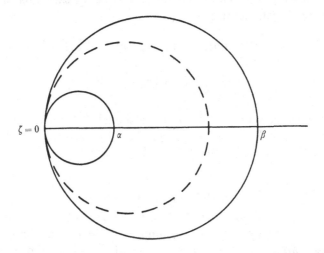

Fig. 9.7.

and $\mathscr{E}_2(\mathscr{G})$. All the prime ends in $\mathscr{E}_1(\mathscr{G})$ have the origin for their impression while the impressions of prime ends in $\mathscr{E}_2(\mathscr{G})$ are segments of circles belonging to the family of mutually tangential circles through the origin, which they have as a common point. These segments are symmetric about the real axis. The accessible points of FR \mathscr{G} are: (i) the origin, accessible along the family of circles corresponding to the ordinates through its pre-images on CD and $C'D'$ (which all belong to $\mathscr{E}_1(B_\infty)$); and (ii) the images in the crescent of the accessible points of FR B_∞ inside the rectangle $CDD'C'$ (which all belong to $\mathscr{E}_2(B_\infty)$). Denote the set of accessible points of FR \mathscr{G} by S and the origin $\zeta = 0$ by O and consider the

set $S-(O)$. Any two points of $S-(O)$ are separated by one of the circles through O whose pre-image is an ordinate through an accessible point on CD and which does not meet $S-(O)$. This follows at once from the symmetry of the figure and the fact that $\mathscr{E}_1(\mathscr{G})$, like $\mathscr{E}_1(B_\infty)$, is residual. The set $S-(O)$ is thus *totally disconnected*. The set S is, however, connected. By a well-known theorem in topology, a set which becomes totally disconnected on the removal of a single point contains no continuum.†

Now suppose that $f(z)$ maps the unit disc conformally on \mathscr{G}. Since every prime end of \mathscr{G} belongs to $\mathscr{E}_1 \cup \mathscr{E}_2$, the radial limit $\lim_{r \to 1} f(r e^{i\theta})$ exists for all $e^{i\theta} \in K$ and

$$\Gamma_P(f, e^{i\theta}) = \lim_{r \to 1} f(r e^{i\theta}) = C_\rho(f, e^{i\theta}) = C_{\mathscr{A}}(f, e^{i\theta}).$$

Therefore $$S = \bigcup_{0 \leqslant \theta < 2\pi} \Gamma_P(f, e^{i\theta}) = \Gamma_P(f),$$

and this set contains no continuum although $f(z)$ is bounded and every point of K is a point of $F(f)$, the set of Fatou points. This proves the assertion which follows Theorem 6.3.

11. The set of asymmetrical prime ends

The subsidiary points in the impression $I(P)$ of a prime end P do not necessarily form a continuum. Let $P = P(\theta)$ under a conformal mapping of $|z| < 1$ by $f(z)$. Since $\Pi(P)$, the set of principal points, is the outer angular cluster set $C_{\mathscr{A}}(f, e^{i\theta})$ the subsidiary points, for $P \in \mathscr{E}_2 \cup \mathscr{E}_4$, are contained in the cluster sets at $e^{i\theta}$ outside a Stolz angle at $e^{i\theta}$. They thus constitute one or two continua. The union of the set $\Pi(P)$ with the set of subsidiary points in the cluster set to the right or left of the Stolz angle we call the *right or left wing* of the prime end. We defined the right and left-hand cluster sets $C_R(f, e^{i\theta})$ and $C_L(f, e^{i\theta})$ in Chapter 4, § 6. These plainly define the right and left wings of the prime end $P(\theta)$ under the mapping. We shall say that a prime end is *asymmetrical* if its right and left wings are not identical; otherwise it is symmetrical. The prime ends $P_1 \in \mathscr{E}_2$ and $P_2 \in \mathscr{E}_4$ illustrated in figure 9.1 are asymmetrical.

† See, for example, Menger [1], p. 207.

As an immediate consequence of the corollary to Theorem 4.11, we now have

THEOREM 9.13. *The asymmetrical prime ends of any simply connected domain form a set which is at most countable.*

For, by that corollary (valid for an arbitrary function), $C_R(f, e^{i\theta}) = C_L(f, e^{i\theta})$ for all except perhaps a countable set of points $e^{i\theta} \in K$, and, since the correspondence between points of K and the prime ends $P(\theta)$ under a conformal mapping by $f(z)$ is one-to-one, the theorem follows.

It is easy to construct domains having an infinite countable set of asymmetrical prime ends, and this shows that Theorem 9.13 is best possible.

We can carry the classification further by partitioning the set of asymmetric prime ends of a domain D into three distinct classes as follows: a prime end belongs to \mathscr{A}_R if its left wing is a proper subset of its right wing, to \mathscr{A}_L if its right wing is a proper subset of its left wing and to \mathscr{A}_{RL} if neither of its wings is a subset of the other.

The problem of finding necessary and sufficient conditions on the decomposition of the circumference K into subsets to correspond under some conformal mapping to the four kinds of prime ends \mathscr{E}_ν ($\nu = 1, 2, 3, 4$) is still unsolved, although solutions have been found in restricted cases. For a homeomorphism which is not required to be conformal the complete solution has been given by Piranian [1].

We conclude by stating without proof the following theorem† on the correspondence of asymmetrical prime ends:

THEOREM 9.14. *Let E_R, E_L, and R_{RL} be three disjoint sets on $|z| = 1$, each at most enumerable. Then there exists a function $f(z)$ mapping $|z| < 1$ conformally onto some domain D in such a way that each point in E_R, E_L, or E_{RL} corresponds to a prime end in \mathscr{A}_R, \mathscr{A}_L or \mathscr{A}_{RL} respectively, while all other points of $|z| = 1$ correspond to symmetric prime ends of D.*

† Collingwood and Piranian [2].

BIBLIOGRAPHY

The abbreviations and form of the references follow, wherever possible, the style used by *Mathematical Reviews*.

ALEXANDROFF, P. S. and HOPF, H. [1]. *Topologie.* Springer-Verlag, Berlin, 1935.

ARMENDÁRIZ, A. A. [1]. *Cluster sets and quasiconformal mappings.* Dissertation, Rice Univ., Houston, 1961.

ARSOVE, M. [1]. The Lusin–Privalov theorem for subharmonic functions. *Proc. London Math. Soc.* (3) **14** (1964), 260–270.

BAGEMIHL, F. [1]. Curvilinear cluster sets of arbitrary functions. *Proc. Nat. Acad. Sci. U.S.A.* **41** (1955), 379–382.

[2]. Ambiguous points of a function harmonic inside a sphere. *Michigan Math. J.* **4** (1957), 153–154.

[3]. Some identity and uniqueness theorems for normal meromorphic functions. *Ann. Acad. Sci. Fenn. Ser. A I*, no. 299 (1961), 6 pp.

[4]. Characterization of the sets $\Phi(f, e^{i\theta})$ and $\Phi(f)$ for a function $f(z)$ meromorphic in the unit disk. *Math. Z.* **80** (1962), 230–238.

[5]. The Lindelöf theorem and the real and imaginary paths of normal functions. *Michigan Math. J.* **9** (1962), 15–20.

[6]. Characterization of the set of values approached by a meromorphic function on sequences of Jordan curves. *Ann. Acad. Sci. Fenn. Ser. A I*, no. 328 (1963), 14 pp.

[7]. Some approximation theorems for normal functions. *Ann. Acad. Sci. Fenn. Ser. A I*, no. 335 (1963), 3 pp.

[8]. Some boundary properties of normal functions bounded on nontangential arcs. *Arch. Math.* **14** (1963), 399–406.

[9]. Analytic continuation and the Schwarz reflection principle. *Proc. Nat. Acad. Sci. U.S.A.* **51** (1964), 378–380.

[10]. Sufficient conditions for the validity of Cauchy's integral formula. *Proc. Nat. Acad. Sci. U.S.A.* **51** (1964), 380–382.

[11]. Meier points of holomorphic functions. *Math. Ann.* **155** (1964), 422–424.

[12]. Sets of asymptotic values of positive linear measure. *Ann. Acad. Sci. Fenn. Ser. A I*, no. 373 (1965), 5 pp.

[13]. *Ambiguous points of arbitrary plane sets and functions.* MRC Technical Summary Report no. 534, Army Research Centre, Madison, Wisconsin. (1965), 24 pp.

[14]. Horocyclic boundary properties of meromorphic functions. *Ann. Acad. Sci. Fenn. Ser. A I*, no. 385 (1966), 17 pp.

BAGEMIHL, F. and ERDÖS, P. [1]. A problem concerning the zeros of a certain kind of holomorphic function in the unit disk. *J. Reine Angew. Math.* **214/215** (1964), 340–344.

BAGEMIHL, F. and PIRANIAN, G. [1]. Boundary functions for functions defined in a disk. *Michigan Math. J.* **8** (1961), 201–207.

BAGEMIHL, F., PIRANIAN, G. and YOUNG, G. S. [1]. Intersections of cluster sets. *Bul. Inst. Politehn. Iaşi* (N.S.) **5** (9), (1959), no. 3–4, 29–34.

BAGEMIHL, F. and SEIDEL, W. [1]. A general principle involving Baire category, with applications to function theory and other fields. *Proc. Nat. Acad. Sci. U.S.A.* **39** (1953), 1068–1075.

[1a]. Spiral and other asymptotic paths, and paths of complete indetermination, of analytic and meromorphic functions. *Proc. Nat. Acad. Sci. U.S.A.* **39** (1953), 1251–1258.

[2]. Some boundary properties of analytic functions. *Math. Z.* **61** (1954), 186–199.

[3]. Valeurs évitées, mais non asymptotiques, pour les fonctions holomorphes dans le cercle-unité. *C.R. Acad. Sci. Paris* **241** (1955), 1195–1196.

[4]. Functions of bounded characteristic with prescribed ambiguous points. *Michigan Math. J.* **3** (1955–56), 77–81.

[5]. Sequential and continuous limits of meromorphic functions. *Ann. Acad. Sci. Fenn. Ser A I*, no. 280 (1960), 17 pp.

[6]. Behavior of meromorphic functions on boundary paths, with applications to normal functions. *Arch. Math.* **11** (1960), 263–269.

[7]. Koebe arcs and Fatou points of normal functions. *Comment. Math. Helv.* **36** (1961), 9–18.

BANACH, S. [1]. *Théorie des opérations linéaires.* Monografje Matematyczne, Warsaw, 1932.

BESICOVITCH, A. S. [1]. On sufficient conditions for a function to be analytic, and on the behaviour of analytic functions in the neighbourhood of non-isolated singular points. *Proc. London Math. Soc.* (2) **32** (1932), 1–9.

BEURLING, A. [1]. Ensembles exceptionnels. *Acta Math.* **72** (1940), 1–13.

BLUMBERG, H. [1]. A theorem on arbitrary functions of two variables with applications. *Fund. Math.* **16** (1930), 17–24.

BONFERRONI, C. [1]. Una proprietà generale delle funzioni, *Boll. Un. Mat. Ital.* (3) **8** (1953), 384–390.

[2]. Alcune proprietà generali di un insieme variabile. *Boll. Un. Mat. Ital.* (3) **9** (1954), 5–15.

BOURBAKI, N. [1]. *Topologie générale,* Livre III, chaps. 1–2, no. 858, Hermann, Paris, 1939.

BRELOT, M. [1]. *Lectures on Potential Theory.* Tata Inst. of Fund. Research, Bombay, 1960.

[2]. Introduction axiomatique de l'effilement. *Ann. Mat. Pura Appl.* (4) **57** (1962), 77–95.

BRIOT, C. and BOUQUET, C. [1]. *Théorie des fonctions doublement périodiques et, en particulier, des fonctions elliptiques.* Paris, 1859.

BROWDER, F. E. [1]. On the proof of Mergelyan's approximation theorem. *Amer. Math. Monthly* **67** (1960), 442–444.

CANTOR, D. G. [1]. A simple construction of analytic functions without radial limits. *Proc. Amer. Math. Soc.* **15** (1964), 335–336.

CARATHÉODORY, C. [1]. Über die gegenseitige Beziehung der Ränder bei der Abbildung des Innern einer Jordanschen Kurve auf einen Kreis. *Math. Ann.* **73** (1913), 305–320.

[2]. Über die Begrenzung einfachzusammenhängender Gebiete. *Math. Ann.* **73** (1913), 323–370.

[3]. *Vorlesungen über reelle Funktionen*, 2nd ed. B. G. Teubner, Leipzig, 1927.

[4]. Zum Schwarzschen Spiegelungsprinzip. *Comment. Math. Helv.* **19** (1946), 263–278.

[5]. *Theory of Functions of a Complex Variable*, Vol. 2, Chelsea, New York, 1954.

CARGO, G. T. [1]. The radial images of Blaschke products. *J. London Math. Soc.* **36** (1961), 424–430.

[2]. A theorem concerning plane point sets with an application to function theory. *J. London Math. Soc.* **37** (1962), 169–175.

[3]. The boundary behavior of Blaschke products. *J. Math. Anal. Appl.* **5** (1962), 1–16.

[4]. Angular and tangential limits of Blaschke products and their successive derivatives. *Canad. J. Math.* **14** (1962), 334–348.

[5]. A connection between Tauberian theorems and normal functions. *Bull. Amer. Math. Soc.* **68** (1962), 400–401.

[6]. Radial and angular limits of meromorphic functions. *Canad. J. Math.* **15** (1963), 471–474.

[7]. Normal functions, the Montel property, and interpolation in H^∞. *Michigan Math. J.* **10** (1963), 141–146.

CARLESON, L. [1]. A remark on Picard's theorem. *Bull. Amer. Math. Soc.* **67** (1961), 142–144.

[2]. Mergelyan's theorem on uniform polynomial approximation. *Math. Scand.* **15** (1965), 167–175.

CARTWRIGHT, M. L. [1]. On the asymptotic values of functions with a non-enumerable set of essential singularities. *J. London Math. Soc.* **11** (1936), 303–306.

[2]. The exceptional values of functions with a non-enumerable set of essential singularities. *Quart. J. Math.* **8** (1937), 303–307.

CARTWRIGHT, M. L. and COLLINGWOOD, E. F. [1]. The radial limits of functions meromorphic in a circular disc. *Math. Z.* **76** (1961), 404–410.

CASORATI, F. [1]. *Teorica delle funzioni di variabile complesse.* Pavia, 1868.

CESARI, L. [1]. Surface area. *Ann. Math. Stud.* no. 35, pp. 298–316. Princeton, 1956.

CHOQUET, G. [1]. Structures des domaines plans et accessibilité. *C.R. Acad. Sci. Paris* **216** (1943), 279–280.

[2]. Topologie de la représentation conforme. *C.R. Acad. Sci. Paris* **216** (1943), 330–331.

[3]. Représentation conforme et topologie. *C.R. Acad. Sci. Paris* **216** (1943), 402–404.

CHURCH, P. T. [1]. Ambiguous points of a function homeomorphic inside a sphere. *Michigan Math. J.* **4** (1957), 155–156.

[2]. Global boundary behavior of meromorphic functions. *Acta Math.* **105** (1961), 49–62.

[3]. Cluster sets of functions on an n-ball. *Canad. J. Math.* **15** (1963), 112–120.

[4]. Boundary images of meromorphic functions. *Trans. Amer. Math. Soc.* **110** (1964), 52–78.

COLLINGWOOD, E. F. [1]. On the linear and angular cluster sets of functions meromorphic in the unit circle. *Acta Math.* **91** (1954), 165–185.

[2]. Sur les ensembles d'accumulation radiaux et angulaires des fonctions analytiques. *C.R. Acad. Sci, Paris* **238** (1954), 1769–1771.

[3]. Sur les ensembles d'indétermination maximum des fonctions analytiques. *C.R. Acad. Sci. Paris* **240** (1955), 1604–1606.

[4]. Sur le comportement à la frontière, d'une fonction méromorphe dans le cercle unité. *C.R. Acad. Sci. Paris* **240** (1955), 1502–1504.

[5]. A theorem on certain classes of singularities defined by cluster sets. *J. London Math. Soc.* **30** (1955), 422–424.

[6]. On a theorem of Eggleston concerning cluster sets. *J. London Math. Soc.* **30** (1955), 425–428.

[7]. A theorem on prime ends. *J. London Math. Soc.* **31** (1956), 344–349.

[8]. On sets of maximum indetermination of analytic functions. *Math. Z.* **67** (1957), 377–396.

[9]. Addendum: On sets of maximum indetermination of analytic functions. *Math. Z.* **68** (1958), 498–499.

[10]. Cluster sets and prime ends. *Ann. Acad. Sci. Fenn. Ser. A I*, no. 250/6 (1958), 12 pp.

[11]. Cluster sets of arbitrary functions. *Proc. Nat. Acad. Sci. U.S.A.* **46** (1960), 1236–1242.

[12]. On functions meromorphic in the unit disc and restricted on a spiral to the boundary. *J. Indian Math. Soc.* (N.S.) **24** (1960), 223–229.

[13]. Cluster set theorems for arbitrary functions with applications to function theory. *Ann. Acad. Sci. Fenn. Ser. A I*, no. 336/8 (1963), 15 pp.

COLLINGWOOD, E. F. and CARTWRIGHT, M. L. [1]. Boundary theorems for a function meromorphic in the unit circle. *Acta Math.* **87** (1952), 83–146.

COLLINGWOOD, E. F. and LOHWATER, A. J. [1]. Inégalités relatives aux défauts d'une fonction méromorphe dans le cercle-unité. *C.R. Acad. Sci. Paris* **242** (1956), 1255–1257.

[2]. Applications of the theory of cluster sets to a class of meromorphic functions. *Proc. Camb. Phil. Soc.* **53** (Part 1), (1957), 93–105.

COLLINGWOOD, E. F. and PIRANIAN, G. [1]. The structure and distribution of prime ends. *Arch. Math.* **10** (1959), 379–386.

[2]. Asymmetric prime ends. *Math. Ann.* **144** (1961), 59–63.

[3]. Tsuji functions with segments of Julia. *Math. Z.* **84** (1964), 246–253.

[4]. The mapping theorems of Carathéodory and Lindelöf. *J. Math. Pures Appl.* (9) **43** (1964), 187–199.

CONSTANTINESCU, C. [1]. Über die Klassifikation der Riemannschen Flächen. *Acta Math.* **102** (1959), 47–78.

[2]. Dirichletsche Abbildungen. *Nagoya Math. J.* **20** (1962), 75–89.

CONSTANTINESCU, C. and CORNEA, A. [1]. Comportement des transformations analytiques des surfaces de Riemann sur la frontière de Martin. *C.R. Acad. Sci. Paris* **249** (1959), 355–357.

[2]. Über das Verhalten der analytischen Abbildungen Riemannscher Flächen auf dem idealen Rand von Martin. *Nagoya Math. J.* **17** (1960), 1–87.

[3]. Spaces of harmonic functions on Riemann surfaces. *Rev. Math. Pures Appl.* **6** (1961), 373–390.

[4]. La théorème de Beurling et la frontière idéale de Kuramochi. *C.R. Acad. Sci. Paris* **254** (1962), 1732–1734.

DENJOY, A. [1]. Sur les fonctions analytiques uniformes qui restent continues sur un ensemble parfait discontinu de singularités. *C.R. Acad. Sci. Paris* **148** (1909), 1152–1153.

[2]. Sur les fonctions analytiques uniformes à singularités discontinues. *C.R. Acad. Sci. Paris* **149** (1909), 258.

[3]. Sur les singularités discontinues des fonctions uniformes. *C.R. Acad. Sci. Paris* **149** (1909), 386.

[4]. Sur les fonctions analytiques uniformes à singularités discontinues non isolées. *C.R. Acad. Sci. Paris* **150** (1910), 32.

[5]. Sur la continuité des fonctions analytiques singulières. *Bull. Soc. Math. France* **60** (1932), 27–105.

[6]. *Articles et Mémoires*, vol. II, pp. 1013–1016, Paris, 1955.

DOLZHENKO, E. P. [1]. On boundary uniqueness theorems and the behaviour of analytic functions near the boundary. *Dokl. Akad. Nauk. SSSR* **129** (1959), 23–26. (Russian)

[2]. On approximation on closed regions and on null-sets. *Dokl. Akad. Nauk SSSR* **143** (1962), 771–774. (Russian)

[3]. The removability of singularities of analytic functions. *Uspehi Mat. Nauk.* **18** (1963), no. 4 (112), 135–142. (Russian)

DOOB, J. L. [1]. On a theorem of Gross and Iversen. *Ann. of Math.* (2) **33** (1932), 753–757.

[2]. The boundary values of analytic functions. *Trans. Amer. Math. Soc.* **34** (1932), 153–170.

[3]. The boundary values of analytic functions. II. *Trans. Amer. Math. Soc.* **35** (1933), 418–451.

[4]. The ranges of analytic functions. *Ann. of Math.* (2) **36** (1935), 117–126.

[5]. Conditional Brownian motion and the boundary limits of harmonic functions. *Bull. Soc. Math. France.* **85** (1957), 431–458.

[6]. Boundary limit theorems for a half-space. *J. Math. Pures Appl.* (9) **37** (1958), 385–392.

[7]. A non-probabilistic proof of the relative Fatou theorem. *Ann. Inst. Fourier (Grenoble)* **9** (1959), 293–300.

[8]. A relativized Fatou theorem. *Proc. Nat. Acad. Sci. U.S.A.* **45** (1959), 215–222.

[9]. A relative limit theorem for parabolic functions. *Trans. 2nd Prague Conf. on Information Theory*, 1960, pp. 61–70, Academic Press, New York, 1961.

[10]. Relative limit theorems in analysis. *J. Analyse Math.* **8** (1960/61), 289–306.

[11]. Conformally invariant cluster value theory. *Illinois J. Math.* **5** (1961), 521–549.

[12]. Boundary properties of functions with finite Dirichlet integrals. *Ann. Inst. Fourier (Grenoble)* **12** (1962), 573–621.

[13]. One-sided cluster value theorems. *Proc. London Math. Soc.* (3) **13** (1963), 461–470.

[14]. Cluster values of sequences of analytic functions. *Sankhyā* Ser. A **25** (1963), 137–148.

ERDÖS, P. and PIRANIAN, G. [1]. Restricted cluster sets. *Math. Nachr.* **22** (1960), 155–158.

FATOU, P. [1]. Séries trigonométriques et séries de Taylor. *Acta Math.* **30** (1906), 335–400.

FAUST, SISTER C. M. [1]. On the boundary behavior of holomorphic functions in the unit disk. *Nagoya Math. J.* **20** (1962), 95–103.

FERRAND, J. [1]. Étude de la représentation conforme au voisinage de la frontière. *Ann. Sci. École Norm. Sup.* (3) **59** (1942), 43–106.

[2]. Étude de la correspondance entre les frontières dans la représentation conforme. *Bull. Soc. Math. France* **70** (1942), 143–174.

[3]. *Représentation conforme et Transformations à Intégrale de Dirichlet bornée.* Gauthier-Villars, Paris, 1955.

FLETT, T. M. [1]. On a theorem of Lindelöf concerning prime ends. *Tohoku Math. J.* (2) **8** (1956), 273–274.

FREUDENTHAL, H. [1]. Enden und Primenden. *Fund. Math.* **39** (1952), 189–210.

FROSTMAN, O. [1]. Potentiel d'équilibre et capacité des ensembles avec quelques applications à la théorie des fonctions. *Medd. Lunds Mat. Sem.* **3** (1935), 1–118.

[2]. Sur les produits de Blaschke. *K. Fysiogr. Sällsk. Lund Förh.* **12**, no. 15 (1939), 1–14.

GÁL, I. S. [1]. On the continuity and limiting values of functions. *Trans. Amer. Math. Soc.* **86** (1957), 321–334.

GALBRAITH, A. S., SEIDEL, W. and WALSH, J. L. [1]. On the growth of derivatives of functions omitting two values. *Trans. Amer. Math. Soc.* **67** (1949), 320–326.

GAVRILOV, V. I. [1]. A theorem of Collingwood and Valiron. *Dokl. Akad. Nauk BSSR* **4** (1960), 450–453. (Russian)

[2]. Distortion in quasiconformal mappings. *Dokl. Akad. Nauk SSSR* **137** (1961), 1278–1279. (Russian)

[3]. Sequential limits of normal meromorphic functions. *Dokl. Akad. Nauk SSSR* **138** (1961), 16–17. (Russian)

[4]. On the set of angular limiting values of normal meromorphic functions. *Dokl. Akad. Nauk SSSR* **141** (1961), 525–526, (Russian)

[5]. The boundary behaviour of Q-quasiconformal mappings. *Sibirsk. Mat. Ž.* **2** (1961), 650–654. (Russian)

[6]. The cluster set of functions pseudo-analytic in the unit circle. *Dokl. Akad. Nauk SSSR* **148** (1963), 16–19. (Russian)

[7]. Boundary behaviour of functions meromorphic in the unit circle. *Dokl. Akad. Nauk SSSR* **151** (1963), 19–22. (Russian)

[8]. Limits of meromorphic and generalized meromorphic functions along continuous curves and sequences of points in the unit circle. *Vestnik Moskov. Univ. Ser. I Mat. Meh.* **1964**, no. 1, 44–55. (Russian; English summary)

[9]. Limits of meromorphic and generalized meromorphic functions along continuous curves and sequences of points in the unit circle. II. *Vestnik. Moskov. Univ. Ser. I Mat. Meh.* **1964**, no. 2, 30–36. (Russian; English summary)

[10]. Some boundary uniqueness theorems for meromorphic functions. *Uspekhi Mat. Nauk* **20** (1965), no. 6 (126), 59–63. (Russian)

[11]. On holomorphic functions which are bounded on sequences of points. *Sibirsk. Mat. Ž.* **6** (1965), 1227–1233. (Russian)

GEHRING, F. W. [1]. On the radial order of subharmonic functions. *J. Math. Soc. Japan* **9** (1957), 77–79.

[2]. The asymptotic values for analytic functions with bounded characteristic. *Quart. J. Math.* (2) **9** (1958), 282–289.

[3]. The definitions and exceptional sets for quasiconformal mappings. *Ann. Acad. Sci. Fenn. Ser. A I*, no. 281 (1960), 28 pp.

[4]. The Carathéodory convergence theorem for quasiconformal mappings in space. *Ann. Acad. Sci. Fenn. Ser. A I*, no. 336/11 (1963), 21 pp.

GEHRING, F. W. and LOHWATER, A. J. [1]. On the Lindelöf theorem. *Math. Nachr.* **19** (1958), 165–170.

GOLUSIN, G. M. [1]. *Geometrische Funktionentheorie*. VEB Deutscher Verlag der Wissenschaften, Berlin, 1957.

GROSS, W. [1]. Über die Singularitäten analytischer Funktionen. *Monatsh. Math. Phys.* **29** (1918), 3–47.

[2]. Zum Verhalten analytischer Funktionen in der Umgebung singulärer Stellen. *Math. Z.* **2** (1918), 243–294.

[3]. Eine ganze Funktion für die jede komplexe Zahl Konvergenzwert ist. *Math. Ann.* **79** (1918), 201–208.

[4]. Zum Verhalten der konformen Abbildung am Rande. *Math. Z.* **3** (1919), 43–64.

HÄLLSTRÖM, G. AF [1] Über eindeutige analytische Funktionen mit

unendlich viele wesentlichen Singularitäten. *Neuvième Congrès des Mathématiciens Scandinaves à Helsingfors* 1938.

[2]. Über meromorphe Funktionen mit mehrfach zusammenhängenden Existenzgebieten. *Acta Acad. Áboensis, Math. et Phys.* **12**, no. 8 (1939), 100 pp.

HARDY, G. H. [1]. *Divergent Series*. Clarendon Press, Oxford, 1949.

HARTOGS, F. and ROSENTHAL, A. [1]. Über Folgen analytischer Funktionen. *Math. Ann.* **104** (1931), 606–610.

HAUSDORFF, F. [1]. *Mengenlehre*, 3rd ed. Reprinted, Dover, New York, 1944.

HEINS, M. [1]. On the boundary behavior of a conformal map of the open unit disk into a Riemann surface. *J. Math. Mech.* **9** (1960), 573–581.

HOBSON, E. W. [1]. *The Theory of Functions of a Real Variable*. Cambridge, 2nd ed., vol. I (1921), vol. II (1926).

HÖSSJER, G. [1]. Bemerkung über einen Satz von E. Lindelöf. *K. Fysiogr. Sällsk. Lund Förh.* **6**, no. 3 (1936).

HÖSSJER, G. and FROSTMAN, O. [1]. Über die Ausnahmestellen eines Blaschkeproduktes. *K. Fysiogr. Sällsk. Lund Förh.* **3**, no. 16, (1933), 8 pp.

HUNTER, ULYSSES. [1]. An abstract formulation of some theorems on cluster sets. *Proc. Amer. Math. Soc.* **16** (1965), 909–912.

[2]. Essential cluster sets. *Trans. Amer. Math. Soc.* **119** (1965), 380–388.

IKOMA, K. [1]. On a property of the boundary correspondence under quasi-conformal mappings. *Nagoya Math. J.* **16** (1960), 185–188.

IVERSEN, F. [1]. *Recherches sur les fonctions inverses des fonctions méromorphes*, Thesis, Helsinki, 1914.

[2]. Sur quelques propriétés des fonctions monogènes au voisinage d'un point singulier. *Öfvers. Finska Vetensk.-Soc. Förh.* **58**A, no. 25 (1915–16), 1–16.

[3]. Zum Verhalten analytischer Funktionen in Bereichen, deren Rand eine wesentliche Singularität enthält. *Öfvers. Finska Vetensk.-Soc. Förh.* **64**A, no. 4 (1921–22), 1–19,

JARNÍK, V. [1]. Sur les fonctions de deux variables réelles. *Fund. Math.* **27** (1936), 147–150.

KAILA, E. [1]. Über die Ränderzuordnung bei konformer Abbildung von mehrfach zusammenhängenden Gebieten. *Ann. Acad. Sci. Fenn. Ser. A* **55** (1940), no. 9, 63 pp.

KAUFMANN, B. [1]. Über die Berandung ebener und räumlicher Gebiete. *Math. Ann.* **103** (1930), 70–144.

KINNEY, J. R. [1]. Boundary behavior of Blaschke products in the unit circle. *Proc. Amer. Math. Soc.* **12** (1961) 484–488.

[2]. Tangential limits of functions of the class S_α. *Proc. Amer. Math. Soc.* **14** (1963), 68–70.

KLEINER, W. [1]. Démonstration du théorème de Carathéodory par la méthode des points extrémaux. *Ann. Polon. Math.* **11** (1962), 217–224.

KOEBE, P. [1]. Abhandlung zur Theorie der konformen Abbildung. *J. Reine Angew. Math.* **145** (1915), 177–225.

KURAMOCHI, Z. [1]. Correspondence of sets on the boundaries of Riemann surfaces. *Proc. Japan Acad.* **36** (1960), 112–117.

[2]. Mass distribution on the ideal boundaries. *Proc. Japan Acad.* **36** (1960), 118–122.

[3]. On the behaviour of analytic functions on the ideal boundary. I. *Proc. Japan Acad.* **38** (1962), 150–155.

[4]. On the behaviour of analytic functions on the ideal boundary. II, III, IV. *Proc. Japan Acad.* **38** (1962), 188–193, 194–198, 199–203.

KUSUNOKI, Y. and MORI, SHIN'ICHI. [1]. On the harmonic boundary of an open Riemann surface. I. *Japan. J. Math.* **29** (1959), 52–56.

[2]. On the harmonic boundary of an open Riemann surface. II. *Mem. Coll. Sci. Univ. Kyoto Ser. A Math.* **33** (1960/61), 209–223.

LANGE, L. H. [1]. A non-Euclidean analogue to a theorem of H. Milloux and its relationship to a theorem of W. Seidel. *Notices Amer. Math. Soc* **5** (1958), 847–848.

[2]. Sur les cercles de remplissage non-euclidiens. *Ann. Sci. École. Norm. Sup.* (3) **77** (1960), 257–280.

[3]. The existence of non-Euclidean cercles de remplissage in certain sub-sets of the unit disc. *Nagoya Math. J.* **19** (1961), 41–47.

LAPPAN, P. A. JR. [1]. Non-normal sums and products of unbounded normal functions. *Michigan Math. J.* **8** (1961), 187–192.

[2]. On continua which are cluster sets. *Notices Amer. Math. Soc.* **11** (1964), 237.

LAVRENTIEV, M. A. [1]. Sur une classe de représentations continues. *Mat. Sb.* **42** (1935), 407–424.

[2]. *Sur les fonctions d'une variable complexe représentables par des séries de polynomes.* Hermann, Paris, 1936.

LEHTO, O. [1]. The spherical derivative of meromorphic functions in the neighbourhood of an isolated singularity. *Comment. Math. Helv.* **33** (1959), 196–205.

LEHTO, O. and VIRTANEN, K. I. [1]. Boundary behaviour and normal meromorphic functions. *Acta Math.* **97** (1957), 47–65.

LELONG-FERRAND, J. *See* Ferrand, J.

LINDELÖF, E. [1]. Sur un principe général de l'analyse et ses applications à la théorie de la représentation conforme. *Acta Soc. Sci. Fenn.* **46**, no. 4 (1915), 1–35.

LITTLEWOOD, J. E. [1]. Mathematical Notes (4): On a theorem of Fatou. *J. London Math. Soc.* **2** (1927), 172–176.

[2]. *Lectures on the Theory of Functions.* Oxford University Press, 1947.

LOHWATER, A. J. [1]. The boundary values of a class of meromorphic functions. *Duke Math. J.* **19** (1952), 243–252.

[2]. Les valeurs asymptotiques de quelques fonctions méromorphes dans le cercle-unité. *C.R. Acad. Sci. Paris* **237** (1953), 16–18.

[3]. On the Schwarz reflection principle. *Michigan Math. J.* **2** (1953–54), 151–156.

[4]. On the radial limits of analytic functions. *Proc. Amer. Math. Soc.* **6** (1955), 79–83.

[5]. The reflection principle and the distribution of values of functions defined in a circle. *Ann. Acad. Sci. Fenn. Ser. A I*, no. 229 (1956), 18 pp.

[6]. Sur le principe de symétrie et la répartition des valeurs des fonctions analytiques bornées. *C.R. Acad. Sci. Paris* **242** (1956), 2278–2281.

[7]. The boundary behaviour of meromorphic functions. *Ann. Acad. Sci. Fenn. Ser. A I*, no. 250/22 (1958), 6 pp.

[8]. The cluster sets of meromorphic functions. *Treizième Congrès des Mathématiciens Scandinaves, tenu à Helsinki 18–23 août 1957*, pp. 171–177. Mercators Tryckeri, Helsinki, 1958.

[9]. The exceptional values of meromorphic functions. *Colloq. Math.* **7** (1959), 89–93.

[10]. On the theorems of Gross and Iversen. *J. Analyse Math.* **7** (1959–60), 209–221.

LOHWATER, A. J. and PIRANIAN, G. [1]. The boundary behavior of functions analytic in a disk. *Ann. Acad. Sci. Fenn. Ser. A I*, no. 239 (1957), 17 pp.

[2]. The sets of ambiguous points of functions of bounded characteristic. *Michigan Math. J.* **4** (1957), 23–24.

[3]. Conformal mapping of a Jordan region whose boundary has positive two-dimensional measure. *Michigan Math. J.* **1** (1952), 1–4.

LOHWATER, A. J. and SEIDEL, W. [1]. An example in conformal mapping. *Duke Math. J.* **15** (1948), 137–143.

ŁOJASIEWICZ, S. [1]. Une démonstration du théorème de Fatou. *Ann. Soc. Polon. Math.* **22** (1949), 241–244.

LUSIN, N. N. [1]. Sur la représentation conforme. *Bull. Ivanovo-Vozn. Polytech. Inst.* **2** (1919), 77–80.

LUSIN, N. N. and PRIVALOV, I. I. [1]. Sur l'unicité et la multiplicité des fonctions analytiques. *Ann. Sci. École Norm. Sup.* (3) **42** (1925), 143–191.

MacLANE, G. R. [1]. Meromorphic functions with small characteristic and no asymptotic values. *Michigan Math. J.* **8** (1961), 177–185.

[2]. Holomorphic functions, of arbitrarily slow growth, without radial limits. *Michigan Math. J.* **9** (1962), 21–24.

[3]. The geometry of functions holomorphic in the unit circle, of arbitrarily slow growth, which tend to infinity on a sequence of curves approaching the circumference. *Duke Math. J.* **29** (1962), 191–197.

[4]. Asymptotic values of holomorphic functions. *Rice Univ. Studies*, **49**, no. 1 (1963), 83 pp.

MacLANE, G. R. and RYAN, F. B. [1]. On the radial limits of Blaschke products. *Pacific J. Math.* **12** (1962), 993–998.

McMILLAN, J. E. [1]. Asymptotic values of functions holomorphic in the unit disc. *Michigan Math. J.* **12** (1965), 141–154.

[2]. Principal cluster values of continuous functions. *Math. Z.* **91** (1966), 186–197.

[3]. On local asymptotic properties, the asymptotic value sets, and ambiguous properties of functions meromorphic in the open unit disc. *Ann. Acad. Sci. Fenn. Ser. A I* (To appear.)

[4]. On radial limits and uniqueness of holomorphic functions. *Math. Z.* **92** (1966), 321–322.

MARKUSHEVICH, A. I. [1]. *Theory of Analytic Functions*. Moscow, 1950. (Russian); translated as *Theory of Functions of a Complex Variable*, I, II, III, Prentice-Hall, Englewood Cliffs, 1965.

MATHEWS, HARRY T. [1]. Cluster sets at isolated and non-isolated singularities. *Proc. Nat. Acad. Sci. U.S.A.* **53** (1965), 1264–1266.

[2]. A note on Bagemihl's ambiguous point theorem. *Math. Z.* **90** (1965), 138–139.

MATSUMOTO, K. [1]. An extension of a theorem of Mori. *Japan J. Math.* **29** (1959), 57–59.

[2]. Exceptional values of meromorphic functions in a neighbourhood of the set of singularities. *J. Sci. Hiroshima Univ. Ser. A* **24** (1960), 143–153.

[3]. On exceptional values of meromorphic functions with the set of singularities of capacity zero. *Nagoya Math. J.* **18** (1961), 171–191.

[4]. Remark on Lehto's paper. A generalization of Picard's theorem. *Proc. Japan Acad.* **38** (1962), 636–640.

[5]. On some boundary problems in the theory of conformal mappings of Jordan domains. *Nagoya Math. J.* **24** (1964), 129–141.

MAZURKIEWICZ, S. [1]. Über die Definition der Primenden. *Fund. Math.* **26** (1936), 272–279.

[2]. Recherches sur la théorie des bouts premiers. *Fund. Math.* **33** (1945), 177–228.

MEIER, K. [1]. Über die Randwerte meromorpher Funktionen und hinreichende Bedingungen für Regularität von Funktionen einer komplexen Variablen. *Comment. Math. Helv.* **24** (1950), 238–259.

[2]. Über Mengen von Randwerten meromorpher Funktionen. *Comment. Math. Helv.* **30** (1956), 224–233.

[3]. Über die Randwerte der meromorphen Funktionen. *Math. Ann.* **142** (1961), 328–344.

MENGER, KARL. [1]. *Dimensionstheorie*. B. G. Teubner, Leipzig, 1928.

MERGELYAN, S. N. [1]. On a theorem of M. A. Lavrentiev. *Dokl. Akad. Nauk SSSR (N.S.)* **77** (1951), 565–568 (Russian); Amer. Math. Soc. translation, no. 86, 1953, 7 pp.

[2]. Uniform approximations of functions of a complex variable. *Uspekhi Mat. Nauk (N.S.)* **7** (1952), no. 2 (48), 31–122 (Russian); Amer. Math. Soc. translation, no. 101, 1954, 99 pp.

MIODUSZEWSKI, J. [1]. On the domains of indetermination of analytic functions. *Ann. Polon. Math.* 7 (1960), 241–248.

MOORE, R. L. [1]. Concerning triods in the plane and the junction points of plane continua. *Proc. Nat. Acad. Sci. U.S.A.* 14 (1928), 85–88.

MUNROE, M. E. [1]. *Introduction to Measure and Integration.* Addison-Wesley, Cambridge, Mass., 1953.

MYRBERG, LAURI. [1]. Über meromorphe Funktionen auf endlich vielblättrigen Riemannschen Flächen. *Ann. Acad. Sci. Fenn. Ser. A I*, no. 286 (1960), 17 pp.

[2]. Über meromorphe Funktionen auf endlich vielblättrigen Riemannschen Flächen, II. *Ann. Acad. Sci. Fenn. Ser. A I*, no. 301 (1961), 9 pp.

MYRBERG, P. J. [1]. Die Kapazität der singulären Menge der linearen Gruppen. *Ann. Acad. Sci. Fenn. Ser. A I*, no. 10 (1941), 19 pp.

NAKAI, M. [1]. On a problem of Royden on quasiconformal equivalence of Riemann surfaces. *Proc. Japan Acad.* 36 (1960), 33–37.

NEVANLINNA, R. [1]. *Eindeutige analytische Funktionen*, 2te Aufl., Springer-Verlag, Berlin-Göttingen-Heidelberg, 1953.

NEWMAN, M. H. A. [1]. *Elements of the Topology of Plane Sets of Points.* 2nd ed., Cambridge, 1951.

NOSHIRO, K. [1]. Contributions to the theory of meromorphic functions in the unit circle. *J. Fac. Sci. Hokkaido Univ.* 7 (1938), 149–159.

[2]. Contributions to the theory of the singularities of analytic functions. *Japan. J. Math.* 19 (1948), 299–327.

[3]. Cluster sets of functions meromorphic in the unit circle. *Proc. Nat. Acad. Sci. U.S.A.* 41 (1955), 398–401.

[4]. *Cluster Sets.* Springer-Verlag, Berlin–Göttingen–Heidelberg, 1960.

OHTSUKA, M. [1]. On the asymptotic values of functions analytic in a circle. *Trans. Amer. Math. Soc.* 78 (1955); 294–304.

PAINLEVÉ, P. [1]. *Leçons sur la théorie analytique des équations différentielles professées à Stockholm 1895*, Hermann, Paris, 1897.

[2]. Sur les singularités des fonctions et, en particulier, des fonctions définies par les équations différentielles. *C.R. Acad. Sci. Paris* 131 (1900), 489–492.

[3]. *Notice sur les travaux scientifiques*, 1900.

[4]. Observations au sujet de la Communication précédente. *C.R. Acad. Sci. Paris* 148 (1909), 1154.

PFLUGER, A. [1]. *Theorie der Riemannschen Flächen.* Springer-Verlag, Berlin–Göttingen–Heidelberg, 1957.

PICARD, E. [1]. Sur une propriété des fonctions entières. *C.R. Acad. Sci. Paris* 88 (1879).

PIRANIAN, G. [1]. The boundary of a simply connected domain. *Bull. Amer. Math. Soc.* 64 (1958), 45–55.

[2]. On a problem of Lohwater. *Proc. Amer. Math. Soc.* 10 (1959), 415–416.

[3]. The distribution of prime ends. *Michigan Math. J.* 7 (1960), 83–95.

202 BIBLIOGRAPHY

[4]. Jordan domains and absolute convergence of power series. *Michigan Math. J.* **9** (1962), 125–128.

PLESSNER, A. I. [1]. Über das Verhalten analytischer Funktionen am Rande ihres Definitionsbereichs. *J. Reine Angew. Math.* **158** (1927), 219–227.

POMPÉIU, D. [1]. Sur la continuité des fonctions de variables complexes. *Ann. Fac. Sci. Univ. Toulouse* (2) **7** (1905), 264–315.

POTYAGAÏLO, D. B. [1]. On the set of boundary values of meromorphic functions. *Dokl. Akad. Nauk SSSR* **86** (1952), 661–663. (Russian)

PRIVALOV, I. I. [1]. *The Cauchy Integral.* Dissertation, Saratov, 1919. (Russian)

 [2]. *The Boundary Properties of Analytic Functions*, 2nd ed. Moscow–Leningrad, 1950. (Russian)

 [3]. *Randeigenschaften analytischer Funktionen* (translation of [2]), VEB Deutscher Verlag der Wissenschaften, Berlin, 1956.

RIESZ, F. [1]. Über die Randwerte einer analytischen Funktion. *Math. Z.* **18** (1923), 87–95.

RIESZ, F. and RIESZ, M. [1]. Über die Randwerte einer analytischen Funktion, *4. Cong. Scand. Math. Stockholm*, (1916), pp. 27–44.

ROYDEN, H. L. [1]. The equation $\Delta u = Pu$ and the classification of open Riemann surfaces. *Ann. Acad. Sci. Fenn. Ser. A I*, no. 271 (1959), 27 pp.

 [2]. On a class of null-bounded Riemann surfaces. *Comment. Math. Helv.* **34** (1960), 52–66.

 [3]. The boundary values of analytic and harmonic functions. *Math. Z.* **78** (1962), 1–24.

RUDIN, W. [1]. Radial cluster sets of analytic functions. *Bull. Amer. Math. Soc.* **60** (1954), 545.

 [2]. On a problem of Collingwood and Cartwright. *J. London Math. Soc.* **30** (1955), 231–238.

RYAN, FRANK B. [1]. The set of asymptotic values of a bounded holomorphic function. *Duke Math. J.* (To appear)

 [2]. A characterization of the set of asymptotic values of a function holomorphic in the unit disc. *Duke Math. J.* (To appear)

SAKS, S. and ZYGMUND, A. [1]. *Analytic Functions.* Monografie Matematyczne, Warsaw, 1952.

SAMUELSSON, Å. [1]. On the boundary behavior of the derivative of analytic functions. *Ark. Mat.* **4** (1961), 219–230.

SCHLESINGER, E. C. [1]. Conformal invariants and prime ends. *Amer. J. Math.* **80** (1958), 83–102.

SCHNITZER, F. and SEIDEL, W. [1]. On the modulus of unbounded holomorphic functions. *Math. Z.* **88** (1965), 301–308.

SEIDEL, W. [1]. On the cluster values of analytic functions. *Trans. Amer. Math. Soc.* **34** (1932), 1–21.

 [2]. On the distribution of values of bounded analytic functions. *Trans. Amer. Math. Soc.* **36** (1934), 201–226.

 [3]. Holomorphic functions with spiral asymptotic paths. *Nagoya Math. J.* **14** (1959), 159–171.

SELBERG, H. L. [1]. Über die ebenen Punktmengen von der Kapazität Null. *Avh. Norske Vidensk.-Akad. Oslo (Mat.-Naturvid. Kl.)*, (1937), no. 10.

SOMADASA, H. [1]. Blaschke products with zero tangential limits. *J. London Math. Soc.* **41** (1966), 293–303.

SPRINGER, G. [1]. *Introduction to Riemann Surfaces*. Addison-Wesley, Reading, Mass., 1957.

STOÏLOW, S. [1]. *Oeuvre Mathématique*. Éditions de l'Académie de la République Populaire Roumaine, Bucarest, 1964.

STORVICK, D. A. [1]. Relative distances and quasi-conformal mappings. *Nagoya Math. J.* **16** (1960), 111–117.

[2]. Cluster sets of pseudo-meromorphic functions. *Nagoya Math. J.* **18** (1961), 43–51.

[3]. Radial limits of quasiconformal functions. *Nagoya Math. J.* **23** (1963), 199–206.

[4]. A localization principle for a class of analytic functions. *Nagoya Math. J.* **23** (1963), 207–212.

SUVOROV, G. D. [1]. Prime ends of a sequence of plane regions converging to a nucleus. *Mat. Sbornik (N.S.)* **33** (75) (1953), 73–100 (Russian; American Math. Soc. Translations, series 2, vol. 1, 67–93)

SZILÁRD, K. [1]. Über die Grundlagen der Funktionentheorie. *Math. Z.* **26** (1927), 653–671.

[2]. Über die Analoga der ganzen rationalen Funktionen in verallgemeinerten Klassen von Funktionen einer komplexen Veränderlichen, I. *Magyar Tud. Akad. Mat. Kutató Int. Közl.* **6** (1961), 375–380.

[3]. Über die Analoga der ganzen rationalen Funktionen in verallgemeinerten Klassen von Funktionen einer komplexen Veränderlichen, II. *Magyar Tud. Akad. Mat. Kutató Int. Közl.* **7** (1962), 125–135.

[4]. Über die topologische Natur einiger allgemeiner Sätze der Theorie der elliptischen Funktionen. *Magyar Tud. Akad. Mat. Kutató Int. Közl.* **8** (1963), 417–423. (Russian summary)

TANAKA, C. [1]. On functions of class *U*. *Ann. Acad. Sci. Fenn. Ser. A I* no. 317 (1962), 12 pp.

TIETZ, H. [1]. Zur Klassifizierung meromorpher Funktionen auf Riemannschen Flächen. *Math. Ann.* **142** (1960/61), 441–449.

TITCHMARSH, E. C. [1]. *The Theory of Functions*. 2nd ed., Oxford University Press, 1939.

TSUJI, M. [1]. On the theorems of Carathéodory and Lindelöf in the theory of conformal representation. *Japan. J. Math.* **7** (1930).

[2]. Beurling's theorem on exceptional sets. *Tohoku Math. J.* **2** (1950), 113–125.

[3]. A theorem on the boundary behaviour of a meromorphic function in $|z| < 1$. *Comment. Math. Univ. St Paul.* **8** (1960), 53–55.

[4]. *Potential Theory in Modern Function Theory*. Maruzen, Tokyo, 1959.

TUMARKIN, G. C. [1]. Sequences of Blaschke products. *Dokl. Akad. Nauk SSSR* **129** (1959), 40–43. (Russian)

[2]. A sufficient condition for a domain to belong to class *S. Vestnik Leningrad. Univ.* **17** (1962), no. 13, 47–55. (Russian; English summary)

URSELL, H. D. and YOUNG, L. C. [1] Remarks on the theory of prime ends. *Mem. Amer. Math. Soc.* no. 3, 1951, 29 pp.

URYSOHN, P. [1]. Über ein Problem von Herrn C. Carathéodory. *Fund. Math.* **6** (1924), 229–235.

VÄISÄLÄ, J. [1]. On normal quasiconformal functions. *Ann. Acad. Sci. Fenn. Ser. AI*, no. 266 (1959), 33 pp.

VALIRON, G. [1]. Points de Picard et points de Borel des fonctions méromorphes dans un cercle. *Bull. Sci. Math.* (2) **56** (1932), 10–32.

WEIERSTRASS, K. [1]. Zur Theorie der eindeutigen analytischen Functionen. *Abh. Königl. Akad. Wiss.* 1876.

WEISS, MAX L. [1]. Cluster sets of bounded analytic functions from a Banach algebraic viewpoint. *Ann. Acad. Sci. Fenn. Ser. A I*, no. 367 (1965), 14 pp.

WENIAMINOFF, W. [1]. Sur un problème de la représentation conforme de M. Carathéodory. *Rec. Math. Soc. Moscou* **31** (1922), 91–93.

WESTON, J. D. [1]. Some theorems on cluster sets. *J. London Math. Soc.* **33** (1958), 435–441.

WHYBURN, G. T. [1]. *Analytic Topology*. New York, 1942.

[2]. Open mappings on locally compact spaces. *Mem. Amer. Math. Soc.* no. 1, 1950, 24 pp.

[3]. *Topological Analysis*. Princeton, 1958.

WOLF, F. [1]. Ein Eindeutigkeitssatz fur analytische Funktionen. *Math. Ann.* **117** (1940–41), 383.

WONG, YIM-MING [1]. A theorem on points of discontinuity of functions. *J. London Math. Soc.* **40** (1965), 324–325.

WOOLF, W. B. [1]. *Radial cluster sets and the distribution of values of meromorphic functions*. Dissertation, University of Michigan, 1959.

[2]. The boundary behavior of meromorphic functions. *Ann. Acad. Sci. Fenn. Ser. A I*, no. 305 (1961), 11 pp.

YOUNG, G. S. [1]. A generalization of Bagemihl's theorem on ambiguous points. *Michigan Math. J.* **5** (1958), 223–227.

[2]. Types of ambiguous behavior of analytic functions, I. *Michigan Math. J.* **8** (1961), 193–200; II. **10** (1963), 147–149.

[3]. The inversion of Peano continua by analytic functions. *Fund. Math.* **56** (1965), 301–311.

YOUNG, W. H. [1]. On the distribution of right and left at points of discontinuity. *Quart. J. Pure Appl. Math.* **39** (1908), 67–83.

[2]. On the discontinuities of a function of one or more variables. *Proc. London Math. Soc.* (2) **8** (1910), 117–124.

[3]. La symétrie de structure des fonctions de variables réelles. *Bull. Sci. Math.* (2) **52** (1928), 265–280.

YOUNG, W. H. and YOUNG, G. C. [1]. On the internal structure of a set of points in space of any number of dimensions. *Proc. London Math. Soc.* (2) **16** (1917), 337–351.

[2]. On the inherently crystalline structure of a function of any number of variables. *Proc. London Math. Soc.* (2) **17** (1918), 1–16.

ZAKHARYAN, V. S. [1]. Cluster values of a class of functions meromorphic in the circle. *Akad. Nauk Armyan. SSR Dokl.* **35** (1962), 3–11. (Russian)

[2]. Some boundary properties of functions analytic in a circle. *Akad. Nauk Armyan. SSR Dokl.* **38** (1964), 199–206. (Russian)

ZORETTI, L. [1]. Sur les fonctions analytiques uniformes qui possèdent un ensemble parfait discontinu de points singuliers. *J. Math. Pures Appl.* (6) **1** (1905), 1–51.

[2]. *Leçons sur le prolongement analytique.* Gauthier-Villars, Paris, 1911.

ZYGMUND, A. [1]. On a theorem of Littlewood. *Summa Brasil. Math.* **2** (1949), 51–57.

[2]. *Trigonometric Series.* 2nd ed., Cambridge, 1959.

[3]. *Trigonometrical Series.* Monografje Matematyczne, Tom v, Warsaw, 1935.

INDEX OF SYMBOLS

INDEX

210 INDEX

Printed in the United States
By Bookmasters